Biotech's

# DICTIONARY *Of* Animal Husbandry

*Compiled & Edited by*

Dinesh Arora

BIOTECH

2014

BIOTECH BOOKS

DELHI - 110035

Biotech's
# DICTIONARY
Of
# Animal Husbandry

*Compiled & Edited by*
*Dinesh Arora*

---



ISBN 81-7622-122-8

*Published by* : BIOTECH BOOKS
1123/74, Tri Nagar,
Delhi - 110 035
Phone: 27382765
e-mail: biotechbooks@yahoo.co.in

*Showroom* : 4762-63/23, Ansari Road, Darya Ganj,
New Delhi - 110 002
Phone: 23245578, 23244987

*Printed at* : Chawla Offset Printers
New Delhi - 110 052

PRINTED IN INDIA

# PREFACE

This dictionary is an up-to-date, authoritative and comprehensive book on the development and production of domestic livestock. Some of the key features of this book include: recent developments in nutrition and feeding, emphasis on sustainable and integrated systems of production, due attention to the conservation and rational use of the vast stock of breeds and types of domestic animals that contribute to the world's biological diversity; and discussion on different animal species and microlivestock, animal fibres and other products including hides and skins, the sustainable production of forage, and the contribution of animals to farm energy supplies in the form of draught, transport and other power functions.

This is a splendid attempt to present alphabetically the farming practices, terminologies with their meanings or interpretations (glosses) in various animal husbandry contexts. Here, in this magnificent creation, a wide range of integrated farming practices has been discussed. The topics include basic practices, the effect of climate, maintenance of health, nutrition and feeding, reproduction and breeding, conservation of farm animal genetic resources, sustainable livestock production systems; about cattle, buffaloes, sheep, goats, horses, pigs, domesticated birds; milk and milk products, meat and carcass by-products; animal fibres, marking livestock for identification and many other topics of real importance. We need to ensure that animals are free from thirst, hunger and malnutrition. They are free from discomfort, pain, injury and disease. Animals should also be kept free from fear. It should also be ensured that animals can engage in relatively normal patterns of animal behaviour.

Therefore, in every respect, this work can be considered as one of the most authoritative and comprehensive lexicons on the development and production of domestic livestock at present.

■ **a bullock work unit**
the average quantity of work that one bullock can do in eight hours span of time.

■ **abattoir**
animals slaughtering place.

■ **abaxial**
not located on the axis of the body, part or organ.

■ **abdomen**
the spongy part of the rear of the body that lies between the vent and the end of breastbone.

■ **abdominal capacity**
it is the distance between the end of the breast and the point of the pelvic bones.

■ **abduct**
1. to move externally from the median plane.
2. to snatch.

■ **aberrant**
dissimilar from the standard type of a species or genus in one or more characters.

■ **aberration**
deviation from the normal, term used in reference to chromosomal abnormalities.

■ **abiogenesis**
an idea that living things may arise from non-living objects. It is also called *spontaneous generation*. This theory is now no longer credited.

■ **abiotrophy**
1. a degenerative disorder, which cannot be attributed to external causes
2. arid occurring after birth-hereditary instead of congenital.

■ **ablactation**
the state of not producing of milk by mammary glands when an offspring is weaned.

■ **abnormal milk**
a kind of milk, which is not even and homogeneous but blistering, ropy, gristly, watery, slimy, straw coloured,

discoloured or bloody, bad odour or flavoured.

- **abomasums**
  the fourth stomach of ruminants which is lying close to the omasum or third stomach, in which digestion gets completed by the supply of digestive juice.
  It is also called as reed stomach, rennet stomach, true stomach and sometimes called maw.
- **absorption**
  the movement of nutrients through the wall of the tract and capillary walls into the circulatory system.
- **acantholysis**
  the separation of the intercellular connections in the prickle-cell layer of the epidermis.
- **acanthosis**
  a diffused hypertrophy of the prickle-cell layer of the epidermis.
- **acapnia**
  a state of diminished carbon-dioxide in the blood.
- **acariasis**
  a state when man or animal are infested with mites or ticks.

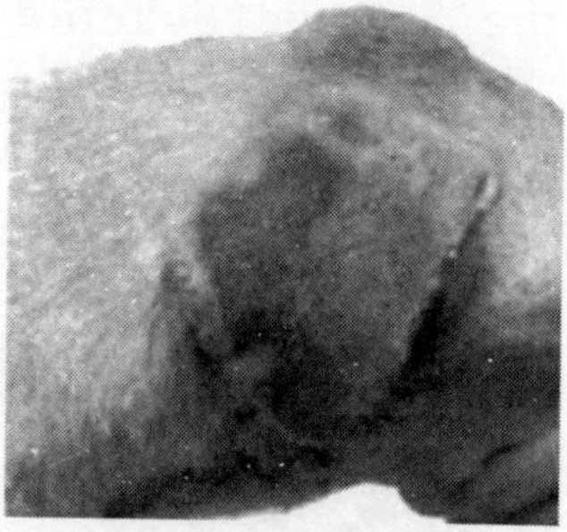

- **abortion**
  the untimely expulsion of foetus or loss of the foetus before the completion of pregnancy often artificial in nature.
- **abortion antigen**
  antigen used in making the agglutination test.
- **abrasion**
  removal of epidermis a little bit resulting in oozing and crusting.
- **abrasive**
  things that can facilitate the mechanical transmission of certain viruses. They tend to increase the number of wounds and also the type of injury conducive to successful establishment of infection.
- **abreast parlour**
  a milking plant structure where the cows are milked

standing side by side.

- **abscess**
temporary collection of pus in a cavity formed by disintegration of tissue.

- **absolute drought**
a situation when a period of at least fifteen consecutive days has not received as much as 0.25 millimetres of rain.

- **acaricide**
a parasiticide used against mites.

- **acarine**
an order of mites and ticks.

- **acarology**
the field of study related to acarina, a large order of small to microscopic arachnids including ticks and mites.

- **acarus**
a parasitic mite which belongs to the natural order acarina.

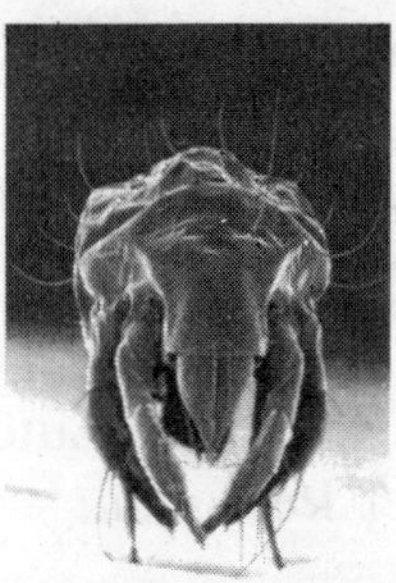

- **accelerator**
a apparatus used to increase the velocity and energy of charged particles, for example, electrons or protons, applying electric and/or magnetic forces.

- **acceptability animal**
the spontaneity with which animals select and eat a forage.

- **accidental error**
an error where it is evenly possible that the sign of the error is plus or minus. In a measurement series, accidental errors are compensative in their total effect.

- **accidents**
injuries which refers to almost every possible form of defection or wounding of the skin, of muscles, tendons, blood-vessels, nerves and internal organs, as well as many varieties of fractures.

- **acclimatisation**
complex of processes of becoming accustomed to a new environmental conditions. It is a series of compensatory alterations in an animal.

- **acclimatisation value**
  the additional value of a flock of sheep acclimatised to a farm over an imported flock that may stray, with loss of sheep (hefted sheep).

- **accredited breeder**
  an accredited breeder's farm is officially accredited by the country authorities.

- **acetonemia (ketosis)**
  a condition characterised by an elevated concentration of ketene bodies in body tissues and fluids. We can find this among high-producing cows in a negative energy balance.

- **acetonuria**
  the presence of acetone in the urine. Usually it is accompanied by acetonaemia.

- **achondroplasia**
  a form of dwarfing due to disease affecting the long bones of the limbs during the prenatal period.

- **achorion**
  a fungus which is only visible through microscope. It causes honeycomb ringworm.

- **achromotrichia**
  the process of greying of the hair.

- **acidosis**
  a disease of animals during the pregnancy period. The symptoms are a metabolic disorder, a decreased alkalinity of blood and tissues due to the presence of excessive amount of acids in the system.

- **acne**
  an inflammation of sebaceous glands or hair follicles. Pustules are also formed with it.

- **acquired character**
  a change or deviation in a character that arises during the course of development of the organism due to the influences of the environmental factors, as opposed to the hereditary factors. This is known as acquired character.

- **acquired immunity**
  a particular immunity which is acquired through exposure to the antigen.

- **ACR**
  abbreviation of Automatic Cluster Removal.

- **acral**
belonging to distant parts of the body, especially the extremities.

- **acromegaly**
a chronic illness caused due to the over function of the anterior pituitary and characterised by an enlargement of the bones and soft parts of the hands, face and feet. Also known as Marie's disease.

- **actinobacillosis**
this is an animal disease mainly of cattle and pigs. It is caused by bacteria (actinobacillus lignieres) resulting the tongue and face getting swelled and stiffened. It is also called as wooden tongue. The disease is also found in sheep and is, sometimes called cruels.

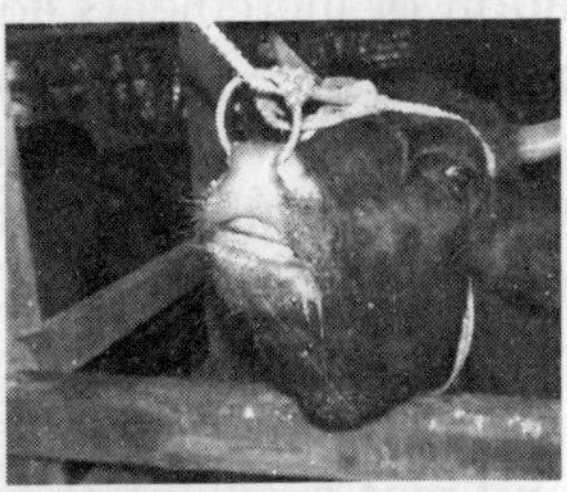

- **actinodermatitis**
a state where inflammation of skin is caused by visible or ultraviolet light.

- **activation analysis**
a method for identification and measurement of chemical elements in a sample of material. First the sample is made radioactive by bombardment with the help of neutrons, charged particles, or gamma rays. The newly formed radioactive atoms in the sample give off characteristic nuclear radiations (such as gamma rays) that can be instrumentally determined. Activation analysis is usually more sensitive than chemical analysis. It is especially useful in animal research, industry, archaeology and criminology.

- **active immunity**
immunity is acquired when an individual produces immune- products (e.g.. antibodies) in response to an antigenic stimulus.

- **active principles**
the material in a drug which are mainly responsible for its effects. For example, the alkaloid morphine is the chief active principle of the crude drug Opium.

- **active transport**
an energy requiring move-

ment of molecules across a membrane.

- **acupuncture**
a centuries-old technique, practised mainly in China, of needle insertion at certain specified spots has been used in veterinary medicine for the purposes of anaesthesia and treatment.

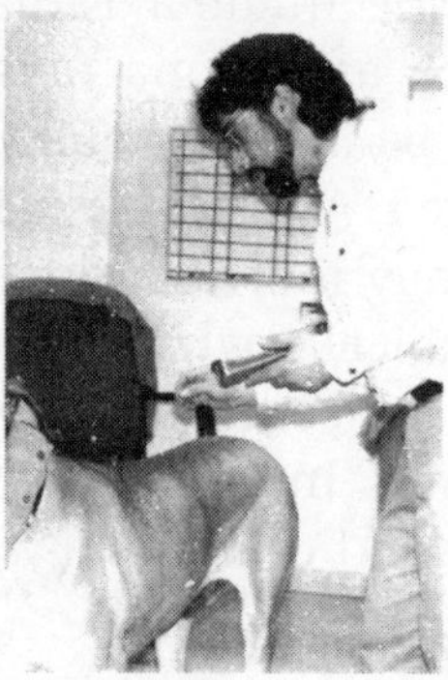

- **acute**
a quality short and severe course.

- **acute disease**
a type of disease that appears rapidly and either causes death quickly or leads to a speedy recovery.

- **ad libitum**
refers to allowing the animal to consume feed on a free-choice basis in animal production.

- **ad lib feeding**
a labour-saving system under which poultry or pigs help themselves to dry meal, etc. and eat as much as they wish. The silage self-feeding to cattle is also considered to be in this category.

- **adaptation**
the way of adjustment of an organism to a new or changing environmental condition.

- **adaptation syndrome**
the defensive response of the body through the endocrine system to systemic injury evoked by stresses and worked out by an initial stage of shock.

- **additive**
1. an ingredient or a combination of ingredients added to the basic feed mixture or parts there have for a specific purpose like to increase feed ingestion or as an aid for digestion or added to produce more desirable consumer products or to alter metabolism.
2. any substance that becomes a component or affects the characteristics of a

feed or food if such substance is not generally recognised, among experts qualified by scientific training and experience to evaluate its safety, as having been adequately shown through scientific procedures to be safe under the conditions of its intended use.

- **additive variance**
  the component of genetic variance due to additive effects of the genes.

- **additivity**
  gene effect that causes a predictable phenotypic change when one gene is substituted for another in its allelic series, irrespective of what other genes are present in the genotype.

- **addled egg**
  a fertile egg in which the embryo died between the seventh and fourteenth day of incubation. We call it a rotten egg generally.

- **adduct**
  bringing toward or moving inward toward the median plane.

- **adenophypophysis**
  interior pituitary.

- **adenoma**
  a non-inflammatory new growth or tumour which looks similar to gland tissue when viewed under the microscope. It is commonly found in positions where gland tissue is not normally present.

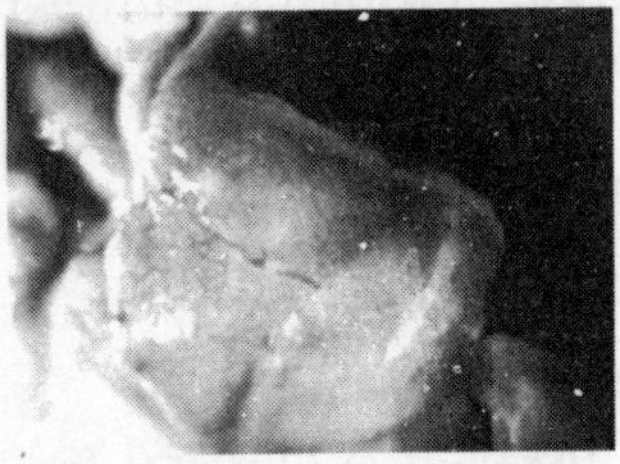

- **adenosine**
  a nucleoside derived from the combination of adenine with a ribose sugar and giving rise to the nucleotide adrenalin acid, one of the nucleic acids.

- **adenosine triphosphate**
  a nucleotide which contains 3 phosphoric acid groups, 2 of which yield a large amount of energy upon hydrolysis.

- **ADF (Acid Detergent Fibre)**
  a fibre that is extracted with acidic detergent in a technique employed to help appraise the quality of forages.

- **ADG (Average Daily Gain)**
the average amount of daily live-weight increment in case of firm animals.

- **adhesions**
it occurs by the uniting or growing together of structures or organs which are normally separate and freely movable, adhesive plasters. These are used to hold the edges of wound together and are frequently employed as dressing.

- **adiabatic change**
if no heat is allowed to enter or leave during the change in pressure and volume of a substance, the change is said to be adiabatic.

- **adipose**
the fatty part of the body.

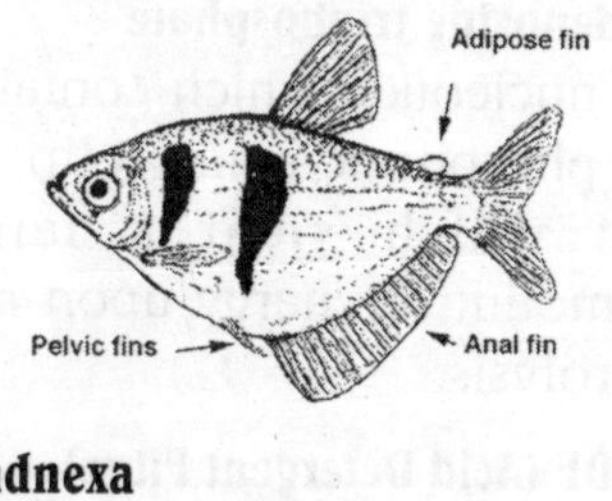


- **adnexa**
the arrangements that are bound to the skin i.e., hair, claws and eccrine, sebaceous and apocrine glands. Extruded hair shafts and claws are more accurately called appendages.

- **adrenal**
close to the kidney.

- **adrenal gland**
ductless gland located upon the upper end of each kidney. Its internal secretion are responsible for increase in blood pressure by contracting the musculature of the blood vessels.

- **adrenalectomy**
the removal of the adrenal gland.

- **adult**
a pullet which has started to lay; a cockerel at nine month's old.

- **adulteration**
to add water to milk.

- **aerobic**
the presence of air or molecular oxygen to live or function.

■ **aftermath**
the post-harvesting recovery growth of forage plants, either by animal or machine.

■ **agalactia**
a contagious disease which is found in goats particularly and sheep less commonly, characterised by inflammatory lesions in the udder, eyes, joints.

■ **agar**
an extract of seaweed that forms a firm gel at temperatures of about 40°c and lower. It is used at a concentration of about 1.5 per cent in microbiological plating media.

■ **age and area**
the areas of distribution of the members of allied species are correlated with their ages as species in any given region.

■ **agglutination titter**
the highest dilution of a serum which causes clumping together of cells, especially bacteria or red blood corpuscles.

■ **agglutinin**
antibody formed against cells (as bacteria). Which, when mixed with these cells, causes them to clump together, or agglutinate.

■ **agistment**
the leading of the grazing rights of a paddock.

■ **air reserves**
the amount of air flow left in band after all the needs of the plant and equipment have been met.

■ **akabane vims**
a possible cause of absortiun in cattle and of birth of abnormal calves.

■ **al daughter**
a female offspring of an artificially inseminated ram.

■ **alanine**
one of the non-essential amino acids.

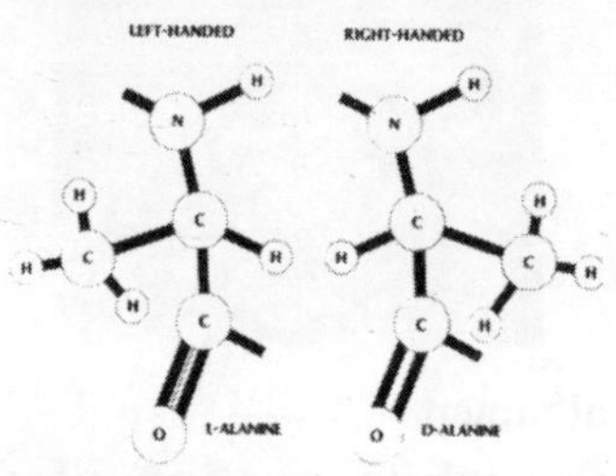


■ **albedo**
albedo is the ratio of the amount of radiation reflected to the amount received on a surface. The

earth's mean albedo is about 35 per cent.

- **albinism**

a condition in which the normal pigment is lacking, such as in the skin, ear and eyes. In plants also lack of normal chlorophyll results in albinism. Albinism is a typical Mendelian recessive character in rat and man.

- **albino**

an animal without having any skin pigment. The reason behind this is the presence of a single recessive gene.

- **albumen**

the white portion of tne reptile's and bird's egg, surrounding the yolk or zygote and supplying food for the embryo.

- **albuminodis**

simple proteins which are soluble in all natural solvent and m dilute acids and alkalies, are coagulated by heat. For example, keratin, collagen, etc.

- **albumins**

proteins closely resembling white of egg and composing in great part all the tissue of the body.

- **albuminuria**

the presence of albumin in the urine, one of the earliest signs of inflammation of the kidneys.

- **alfalfa leaf meal**

the powdered form of dried alfalfa (lucerne) leaves. After collecting the leaves reasonably free from other crop plants or weeds, the meal on an average contain about 20% crude protein and less than 18% crude fibre.

- **alfalfa meal**
  the ground product, which is obtained from the dried alfalfa plant as a whole (stem containing leaves).

- **alfalfa stem meal**
  the ground meal obtained from aerial portion of the panel from which leaves have been excluded.

- **algae**
  chlorophyll-bearing microscopic plants that synthesise food by photosynthesis. They are mostly aquatic but without true stems, roots, or leaves.

- **alimentary**
  related to feed or food.

- **alimentary canal**
  a tube or cavity in any animal concord with digestion and absorption of food. Usually the terminal openings are mouth and anus. The length is about 30 ft. in man.

- **alkaloids**
  an organic nitrogenous compounds occurring naturally in certain plant species, to which is due their marked effects, at morphine and nicotine.

- **allantois**
  a tubular sac that in the initial stage develops from the hind gut of the embryo. It later arches around in order to envelope the embryo completely. Through its development of blood vessels, it performs tile function of interchange between the embryo and egg albumen. In mammals it performs the function of metabolic interchange between embryo and mother. The sac contains a fluid called allantoic.

- **allergy**
  an excessive sensitivity to a common substance namely pollen, many foods, natural oils, hair, dust, feathers, medicines, etc. This sensitivity is shown being manifested in many ways such as hive, rash, asthmatic attacks, higher blood pressure, headaches and nasal congestion. Such symptoms are external manifestation

of antigen-antibody reactions- the antigens being the substances mentioned above and the antibody being the specific antagonistic materials in the tissues and blood plasma.

- **allosome**
  any of the typical chromosomes, especially sex chromosomes.

- **allosteric effectors**
  the small molecules, usually metabolites, that bind of allosteric proteins at a site other than the active site in order to cause a change in protein shape.

- **alopecia**
  baldness, loss or deficiency of hair, natural of abnormal.

- **alpha amino (a-amino)**
  this designates the location of an pHs group on the alpha carbon, as in amino acids.

- **alpha particle**
  a positive charged particle (two protons and two neutrons) emitted by certain radioactive materials, it is the feast penetrating of the three common types of radiation (alpha, beta, gamma) emitted by radioactive material, being stopped by a sheet of paper.

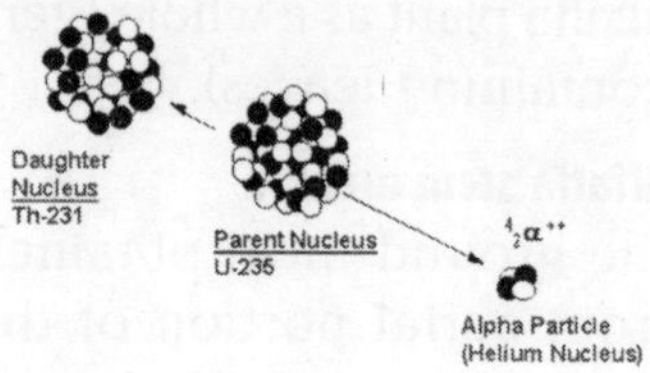


- **alpine**
  a breed of goat often found in Swiss Alps, from which the British Alpine breed has been developed.

- **alternate AM PM sampling plan**
  a dairy record-keeping plan in which milk weights and samples are taken at only one milking. Each test period is the milking alternating from am to pm in succeeding test periods.

- **alveolus**
  an empty follicle of cells. In the case of mammary tissue the cell, lining the lumen of an alveolus manufacture, milk and secrete it into the lumen.

- **amaurosis**
  the deprivation of sight.

- **ambient**
  means which surrounds. For example, the tempera-

ture about us is referred to as ambient temperature.

- **ambient air temperature**
  temperature of the air encompassing an object on all sides.

- **ambient temperature**
  the prevailing or surrounding temperature.

- **amblyopia**
  a diminution in the power of vision.

- **ameiosis**
  the replacement of normal meiosis by single nuclear division so that the chromosome number is not halved.

- **amensalism**
  a form of biological interaction in which one organism is inhibited by another, but the other is in no way affected by it.

- **amitosis**
  a nuclear division which does not involve mitosis. Here the nucleus is divided into two by a constriction. The two nuclei formed from the parent nucleus are usually un-identical.

- **ammoniated**
  a substance which is combined or impregnated with ammonia or an ammonium compound.

- **ammoniated mercury**
  a white poisonous powder which is soluble in acids. It is used as antiseptic and parasiticide, especially in the form of ointment having about five per cent ammoniated mercury.

- **amnion**
  a thin membrane forming a closed sac around the developing embryo. It contains the amriolic fluid, in which the embryo is dipped.

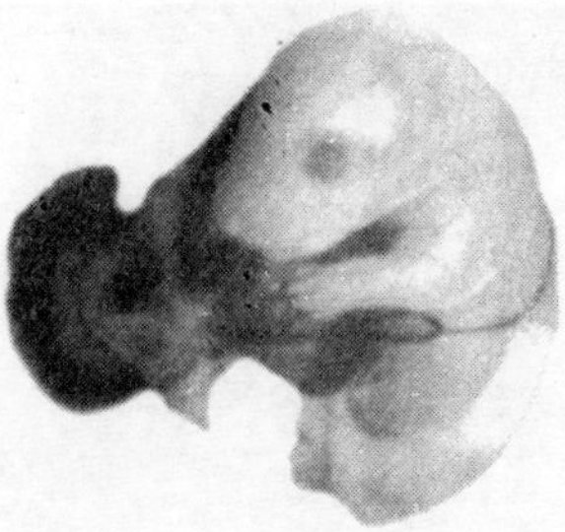

- **amniota**
  any reptile, bird or mammal whose embryos possess an amnion and an allantois.

- **amoeboid cell**
  a kind of cell which is capable for amoeboid movement or capable of forming pseudopodia, in the general body tissues of many lower

phyla and in the circulatory system of higher phyla.

**Amoeboid movement**

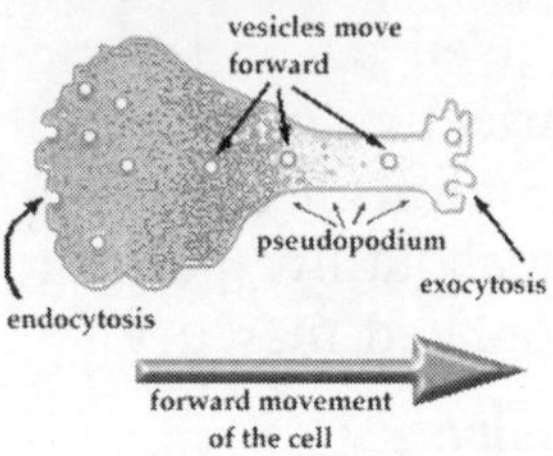


- **amphibian**
an organism (e.g. frog) capable of living either on land or in the water.

- **ampicillin**
a semi-synthetic penicillin that is active against both gram-positive and gram-negative bacteria. These are not resistant to penicillinase, but can be administered by mouth.

- **amplifier**
an electrical mechanism for increasing the current of a pulse of electricity (strengthening electric impulses).

- **AMR**
the shortened form of Automatic Milk Release.

- **anabolism**
a constructive reaction by which simple substances are converted by living cells into more complex compounds (constructive metabolism).

- **anal gland**
the gland that is associated with the rectum or anus in a wide variety of vertebrate with excretory functions.

- **analgesia**
the loss of the power to feel pain without loss of consciousness or of the power of movement of the part although that may, in some instances, be interfered with.

- **analgesic**
drug that lessens or prevents pain. For example, morphia.

- **analogue**
a chemical that resembles another in structure or function.

- **analogy**
the similarity in function, but not in structure or development.

- **analysis**
1. the record of an examination.
2. the resolution of complex problems, bodies, or liquids into their elements.
3. the resolution of anything into its elements.

- **analysis of variance**
the method through which we partition the variations into components assignable to different causes.

- **anamestic**
small variable bones filling spaces between larger bones of more filed position, as in fish skulls.

- **anamnesis**
the past record of a particular patient.

- **anaphase**
the third stage of mitosis. The main characteristic of this phase is the separation of centromere and identical anaphrodisia, impairment of sexual appetite.

- **anaplasmosis**
a serious, infectious blood disease of cattle which is not transmissible to man, but goat, sheep, buffalo, camel etc., are susceptible. It is caused by a minute parasite: anaplasma marginal that invades and destroys large number of red blood cells. The agents of infection are licks, mosquitoes and other biting insects. Symptoms include rise in temperature from 103°F to 107°F and pulse rate become 70 to 140 a minute. The red blood cell count decreases from a normal of about 5 to 7 millions per cent of blood to 2 millions or lees in severe cases. The sick cattle are generally constipated and abortion often takes place in advanced pregnancy. In order to provide proper treatment, sick animals should be kept in the shade and given plenty of water and should be protected against flies and the biting insects. In early cases potassium arsenate solution have been found to be useful.

- **anatomy**
a branch of science which deals with morphology (structure of organism) and

usually but not always, refers to the parts seen without a micro-scope.

■ **anatoxin**
a kind of toxin that remains harmless by heat or chemical means but capable of stimulating the formation of antibodies.

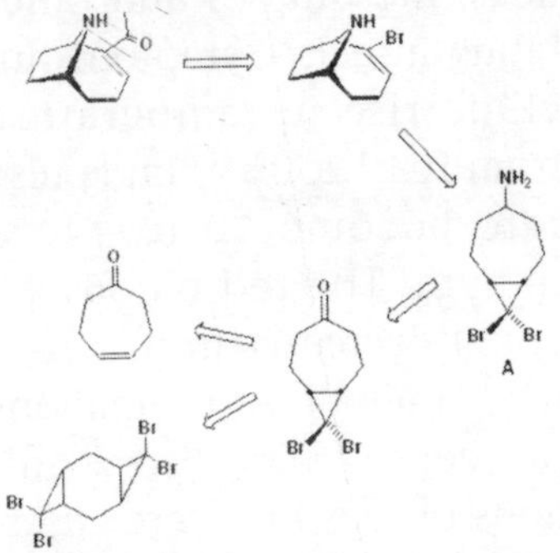


■ **ancestor**
a previous generation animal that has passed on genes through a line of descent.

■ **anaemia**
a physical condition in which the blood is deficient in the amount of needed white blood corpuscles or in the number of red blood corpuscles or in both. The main characteristics are the pains of the skin and mucous membranes, loss of energy and palpitation of the heart, which is felt by the individual.

■ **anaemic**
the shortage of the number of red blood cells.

■ **anaesthesia**
absence of sensitivity or loss of the feeling of pain, touch, cold, or other sensation, produced by ether, chloroform, morphine and other compounds, by hypnotism, or as the result of hysteria, paralysis, like disease.

■ **anaeptrus**
this refers to a period without any sexual desire.

■ **aneurysm**
the term refers to a spindle shaped, globular or cylindrical dilation of a blood veel. It often has a heavy deposit of fibrin causing an interference with the circulation of the blood. Imangloma is a tumour composed of a large number of tissues. These are present in the livers of cattle, when they are
A sharp bend formed by the meeting of two border or lines.

■ **angora goat**
a special kind of goat which originated in Asia minor and

which is now found in large numbers in Texas and South Africa. It produces mohair.

- **angstrom**
  a unit of measurement 1/10,000.000 of a millimetre ($10^{-7}$ mm).

- **animal fats**
  fats derived from animals and characterised by a high proportion of saturated fatty acids, as opposed to the plant fats or oils which are characterised by a high proportion of unsaturated acid.

- **Animal Protein Factor (APF)**
  the same compound which was once recognised as an unidentified growth factor present in animal feeds found absolutely essential for poultry and swine growth. It is now known to be the same as vitamin $B_{12}$.

- **animal unit**
  a mature cow or its equivalent- one animal unit approximates 1 horse, 5 sheep or 2 yearling calves.

- **animal unit month**
  the forage required to support an animal unit 30 days, or 310 tons of hay, or 300 pounds of TDN.

- **animalise**
  the chemical treatment of vegetable fibres, such as cotton on flax in order to resemble the fibre with animal fibres like wool, in their behaviour toward wool dyes.

- **anion**
  an ion that carries a negative charge of electricity. Included the non-metals, the acid radicals and the hydroxyl ion.

- **ankylosis**
  the condition of a joint in which the movement is restricted by union of the bones.

- **anlage**
  first identifiable rudiment of a developing tissue or organ.

- **annular**
  resembling a ring.
- **anodynes**
  any pain-relieving drug.
- **anoestrous pig**
  sow or gill which has failed to show a sexual desire for mating for an indefinite period.
- **anogenital**
  refers to both the anal and the genital area.
- **anomaly**
  a marked deviation from the normal condition.
- **anoxeria**
  lack or loss of appetite for food.
- **anoxia**
  lack of oxygen. If oxygen is not provided with, pilots develop anoxia (altitude sickness) when flying at high altitudes.
- **antacid**
  a substance that counteracts tidily.
- **antagonism**
  1. when one or more antibiotics are acting together produces inhibitory effect which is less than the additive effects of independent antibiotics.
  2. means injury, killing or inhibition of growth of one species of micro-organisms by another,
- **antagonist**
  an organism having the capacity to inhibit the growth of or interfere with the activity of another organism. This is because of neutralising effect of their antibiotic substance.
- **antagonistic medicine**
  any medicine that is able to counteract the action of another medicine.
- **interior**
  the front or forward part. It indicates to the ventral surface of the body in human anatomy.
- **interior pituitary**
  endocrine gland which is located at the base of the brain. It secretes and stores a number of different hormones many of which are involved in reproduction.
- **anthrax**
  an acute and infectious disease of livestock animals. It is mostly acquired through infected food. Cattle, sheep, goats etc., are commonly af-

fected by anthrax. A germ this is one of the most unusual pathogenic microorganisms encountered in nature causes it. Outbreaks frequently follow the not dry summer when the herbage growth is scanty, which necessitates grazing close to the soil the spread of anthrax may be brought about by contaminating the soil, water, pasture or hay plants with the excreta and discharge of diseased animals. The per-acute-unit forms of the disease usually terminates in deaths the body temperature may reach 107° F and rumination ceases.

- **anthrax bacterin**

it tends to stimulate the treated animal to produce immune bodies, i.e. , active immunity. Its anthrax bacterin being sterile, is in itself incapable of producing disease in the treated animal, it is, therefore, safer ihun any of the anthrax-spore vaccines.

- **anthrax spore vaccinations**

living anthrax spores used alone or in combination with anthrax serum. The spores in these vaccines become so weakened that they will not produce the disease in livestock under ordinary conditions, but some susceptible animals may die. For this reason these vaccines should be used only on those premises where the disease has occurred earlier,

- **anthrax vaccination**

a preventive vaccine against a serious disease of livestock called anthrax.

- **anthropophilic**

1. showing a preference for man.
2. pertaining to anthropophilic fungi whose primary host is man. (used in dermatology)

- **anti-anthrax serum**

serum directly responsible for increased resistance to anthrax in direct proportion to the quantity of serum which is given to the animal. It is of value both as a preventive and as a therapeutic agent. It should be used as a preventive only. In case of immediate protection it should be noted that the immunity, which it confers, is of relatively short duration.

- **antibacterial agent**
agents that kill or inhibit the growth of bacteria, e.g., antiseptics, antibiotics and disinfectant.

- **antibiotic**
a compound produced by a micro-organism that inhibits the reproduction, or causes the destruction of other micro-organisms. This term has been extended further now and defined as a compound produced by a micro-organism or a plant, or a close chemical derivative of such a compound that is toxic to micro-organ isms from a number of other species.

- **antibiotic feed supplement**
a feeding material, chiefly used for its antibiotic content, e.g.., terramycin, penicillin and streptomycin. Thus it is a single or a combination of antibiotics having growth-promoting properties. They give rise to a more rapid growth by altering the flora in the digestive tract of the animals.

- **anti blackleg serum**
a serum that is used for treating aches which are already affected with blackleg as well as for preventing it. It is produced by injecting a pure culture of blackleg organisms into the veins and later under the skin of horses and drawing the blind afterwards.

- **antibody**
a protein substance (modified type of blood-serum globulin) developed or synthesised by lymphoid tissue of the body in response to an antigenic stimulus. Each antigen elicits production of a specific antibody to disease defence; the animal must have had an encounter with the pathogen (antigen) before specific antibody is found in its blood.

- **anti-coagulants**
substances that resist the clotting of the blood. They include dicoumarol and hcparin.

- **antidiuretic hormone**
hormone of the posterior pituitary which promotes water resumption in the kidneys and increases blood pressure by constricting arterioles and coronary blood vessels.

■ **antidotic**
a chemical substance which is produced by micro organism and has the capacity to inactivate the injurious effect of bacterial toxins.

■ **antigen**
a substance or complex of high molecular weight, usually composed of protein or protein-polysaccharide complex in nature, which when, foreign to the blood stream of an animal, on gaining access to the tissues of such an animal stimulates the formation of specific antibody and reacts specifically in vivo or in vitro with its homologous antibody.

■ **antigen antibody reaction**
any sort of reaction which takes place between an antigen and its corresponding antibody it may involve agglutination, precipitation, complement-fixation etc. The sensitivity of allergic reactions of many individuals towards certain kinds of foods, pollen, dusts, feathers, hair, drugs, etc. are all examples of antigen-antibody reactions.

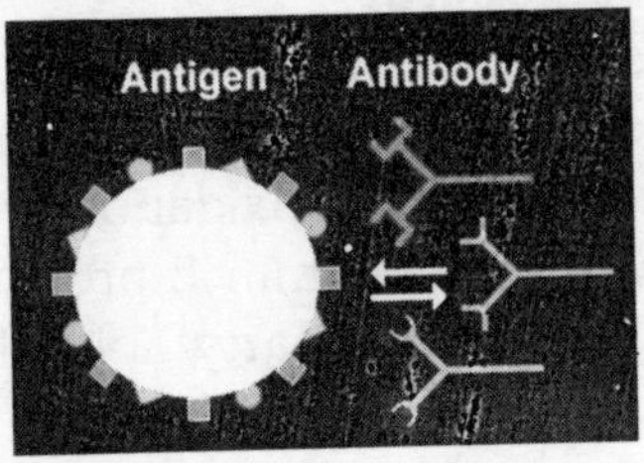


■ **anti-histamines**
organic compounds that, when present in minute amounts, will counteract or resist the action of histamine in animal body tissues where it occurs as the result of allergic reactions.

■ **anti-infective**
a remedy which is administered to overcome an infection in livestock, e. g., sulfonammes, penicillin, etc.

■ **anti-metabolite**
a substance that bears a close structural resemblance to one required for normal physiological functioning and exerting its effect, perhaps, by replacing, or interfering with the utilisation of the essential metabolite.

■ **antimycyn**
an antibiotic which functions as a respiratory, inhibitor in mitochondria, but generally not in bacteria.

- **anti-oxidant**
  substances having the property of protecting other substances from oxidation. For example vitamin E protects unsaturated fatty acids by oxidising themselves.
- **anti-pruritic**
  an agent that is used for relieving itching.
- **anti-pyretics**
  are drugs used to reduce temperature during fevers.
- **anti-rheumatic**
  drug that is used to prevent or cure rheumatism in animals, e.g., sodium saticylale.
- **antiseptic**
  a substance that prevents growth and development of micro-organisms either by destroying hem bactericidal action) or inhibiting their growth (bacteriostatic action). (derived from the Latin anti, meaning against and sepsis, putrefaction)
- **anti-serum**
  a substance that contains an antibody or antibodies. It gives temporary protection against certain specific infectious diseases.
- **anti-sialics**
  substances that reduce salivation, e.g., atropine.
- **anti-spasmodic**
  therapy used to relieve spasm.
- **antitetanic serum**
  a serum used against tetanus. (Nowadays the antitoxin is preferred.)
- **antitoxin**
  an antibody formed against poisonous toxins, such as the bacterial exo-toxins, which specifically neutralises (counteracts) the effects of the toxic diphtheria antitoxin, obtained from the blood of horses infected with diphtheria, is injected into a person to make him immune to diphtheria or to treat him if the person is already infected with the disease.
- **antrum**
  a cavity or chamber.
- **anus**
  posterior opening of the digestive tract. In vertebrates an anus is used only for the elimination of faeces.
- **aob**
  Any Other Breed.

■ **aoc**
Any Other Colour.

■ **aov**
Any Other Variety.

■ **apex**
the top or pointed extremity of a corneal mass— the top angle of a triangle.

■ **apf (Animal Protein Factor)**
the original label given to vitamin $B_{12}$.

■ **aphrodisia**
a sexual excitement especially when there is morbid or excessive release of hormones e.g. , sexual odours released by male insects to excite sexually female insects.

■ **aplasia**
a condition of failure or lack of full development or growth.

■ **apnea**
a condition where breathing is stopped or decreased.

■ **apocrine**
a type of glandular secretion in which the secreted products become concentrated at the free end of the secreting cell and are thrown off, along with the portion of the cell in which they have accumulated, as if the mammary gland.

■ **apodous**
without leg.

■ **appetitive behaviour**
any type of animal behaviour resulting in the satisfaction of a specific need, e.g., searching by an animal for food or a male.

■ **apterous**
wingless.

■ **aquaculture**
the raising of plants or animals, as fish or shellfish, in or under a sea, a lake, a river, or any other body of water.

▪ **arable**
land that is suitable for cultivation; tillable.

▪ **arachidonic**
a 20-carbon unsaturated fatty acid having four double bonds.

▪ **arboreal**
living in or among trees.

▪ **arbovirus**
anyone of several groups of small viruses transmitted by arthropods such as mosquitoes and ticks. Yellow fever, dengue and equine encephalomyelitis are caused by arboviruses.

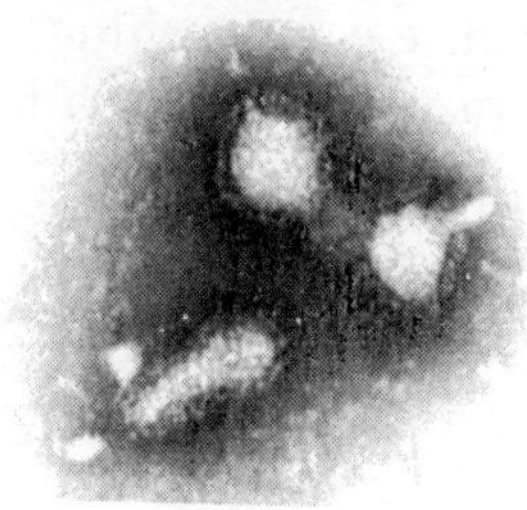

▪ **arch**
the convex curvature of the top line of swine. Generally considered the profile from the ears to the tail head.

▪ **archimedes spiral**
a plane curve generated by a point moving away from or toward a fixed point at a constant rate while the radius vector from the fixed point rotates at a constant rate.

▪ **argente**
a breed of rabbit of French origin, with a silvery coal. There are several shade varieties, e.g., argente blue, argente silver, etc.

▪ **arginine**
one of the essential amino acids.

▪ **arid**
1. a climate or a region where the rainfall is barely sufficient to support vegetation;
2. deficient in rainfall.

▪ **aridity**
the climate characterised by excess of evaporation over precipitation.

▪ **aridity index**
annual water deficit as percentage-of annual potential evapo-transpiration of

moist climates.

- **arsenical poisoning**
poisoning caused due to the intake of any compound containing arsenic. In poultry and livestock, it may occur as a result of eating of grasshopper bait containing sodium arsenic particularly if the bait has not been spread evenly on a range or from eating the poisoned grasshopper arsenical poisoning may also be caused by carelessness in handling arsenical dips or by injudicious administration of arsenical drugs. The poisoning results in trembling, restlessness increased thirst, rapid breathing, colic, prostration, staggering and in acute cases death may occur in 2-3 hours. Treatment involves removal of stomach contents with a stomach tube, give a handful sodium byphosphate (sodium thiosulphate) in a drench and/or inject a solution of the same drug intravenously.

- **arsenilic acid**
one of the organic compounds of arsenic axed as a growth supplement for pigs and poultry and for the control of swine dysentery.

- **arteriosclerosis**
a progressive thickening and hardening of the waffles of the arteries, often associated with high blood pressure of with chronic disease of the kidneys.

- **arteriovenous anastomosis**
small blood vessel that short-circuits bloods from the arterioles directly to the vepules thus by passing the capillary bed. It has in response to stimulation of the sympathetic nerves that supply them.

- **artery**
a blood vessel which carries blood away from the heart.

- **arthritis**
an inflammation of a joint. It results in an enlargement and impaired movement of the joint that causes extreme lameness, often accompanied by fever.

- **arthropoda**
a phylum of the animal kingdom which includes insects, spiders and crustacea; characterised by a coating that selves as an external skeleton and by legs with distinct moveable segments or joints.

- **arthrospore**
  a spore formed by separation of hyphae with subsequent separation of cell.

- **articular**
  related to a joint.

- **artificial brooding**
  the handling of baby chicks without the aid of hen, accomplished by means of heated harder; has advantages over natural brooding. Brooding chicks possible in any time of the year and can be controlled by sanitary conditions, regulation of temperature and planned feeding possible.

- **artificial hatching**
  incubation of eggs by methods other than by natural hatching; normally by using machines called, incubators more advantageous than natural hatching.

- **artificial insemination**
  deposition of sperm in the female genitalia by instruments rather than by male lea. the semen is collected from the male by artificial . The semen is inseminated into the front by placing a portion of it cither in a collected or a diluted form into the cervix or uterus by mechanical methods in the proper time and under most hygienic conditions. Otherwise it may not lead to conception or fertilisation. A bull produces 50- 100 times more semen in a normal mating than is needed for a cow to conceive. Thus, by the collection, dilution and refrigerated storage of a bull's semen, it is possible for that bull to sire, via AI far more progeny during and even after its lifetime than would be possible by normal mating. A.I. also reduces the spread of venereal disease and enables farmers to economise by dispensing with their own or a commercial hull. Diluted semen is stored at — 196°C, after the addition of glycerol and may be kepi for years. A.I. plays an important part in cattle breeding (moist of the dairy cattle) in India i.e. is now also used for steep. Most adult rams can produce from 6000 to 8000 doses of fresh chilled semen, or 2500-5000 doses of deep frozen semen a year.

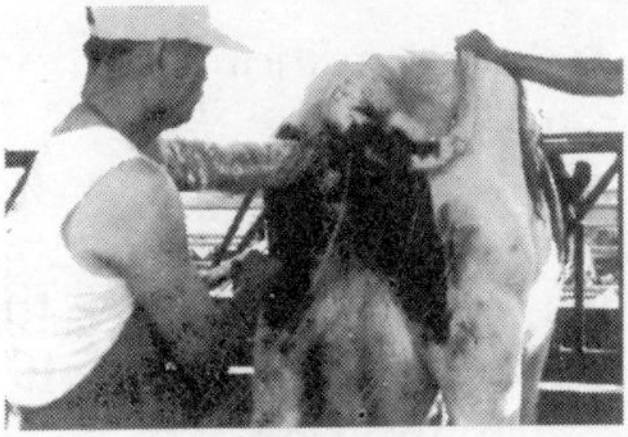

- **artificial vagina**

the artificial vagina is used in order to collect semen from bull artificially instead of real female vagina. It consists of an outer cylinder of heavy rubber casing, glass, ebonite, or metal with a team rubber inner tube, which gets attached to the outer cylinder in such a way that a watertight space gets created between the inner and outer waifs. One end of the inner lube is kept open; the tube are the other end has been tapered to fit tightly over a graduated test tube in which the semen has been collected at the time of ejaculation. There is a small valve in the outer wall for introducing warm water into the space between the inner and outer surfaces. A temperature of 410c has been found to be near optimum for most animal species. The inner tube is well lubricated.

- **artificially dried**

the drying process other than natural means. Dehydrated.

- **artiodactyla**

cattle, swine, goats and deer.

- **arteriole**

a small artery that supplies blood directly to the capillary. It resembles the large arteries except that the tunica media, in all the pulmonary arterioles, is composed entirely of smooth muscle cell, 15-20 m in length.

- **as fed**

feed as it is consumed by an animal.

- **as hatched**

the unsexed day-old chicks packed for sale as they are hatched in the incubators.

- **ascariasis**

injuries caused by the large intestinal round worms or ascarids (large roundworms) which are present in the lumen of the small intestine, the hosts (horses, cattle, swine) obtain their parasites through swallowing their microscopic infective eggs with forage, dry feed, soil or water adult worms may damage the liver and lungs. The adult worms in the intestine take the blood of the host. Occasionally, they even make perforations in the intestines and produce peritonitis. Ascariasis slows down the growth of young host animals, producing more or less permanent stunting and sometimes-even death if the worms invade vital organs. The treatment involves the administration of chenopodium oil. Swine may also be treated with sodium fluoride and horses with carbon disulphide.

- **ascarib**
  large intestinal roundworm.

- **ascarops strongylina**
  a kind of thick stomach-worm species which is occurring in swine.

- **ascorbic acid (Vitamin C)**
  an essential growth element that is found in fruits (especially citrus) and vegetables. It commonly known as vitamin C

- **ascospore**
  the sexual spore of ascomycetes. Born in an ascus.

- **asepsis**
  a septic condition, methods, or treatment.

- **aseptic**
  free from living germs that cause disease, putrefaction or fermentation.

- **aseptic packaging**
  method of filling packages without allowing the entry of micro-organisms. Commonly used for sterile milk products to provide improved keeping quality.

- **asexual**
  sexless.

- **ash**
  1. residue left after burning primarily composed of minerals.
  2. a term used in the analysis of feeding stuffs for the residue left after through burning and the deduction of organic matter. Otherwise called mineral matter.
  3. constituent of milk comprising various sail in solution of which calcium and potassium phosphates and sodium chloride predominate and also various vitamins (a, b, c, d and e). In normal milk the ash proportion remains fairly constant

at about 0.75% (milk composition). A significant reduction may indicate added water.

- **aspartic acid**
one of the non-essential ammo acids.

- **aspect**
1. the seasonal appearance of a community.
2. the direction towards which a slope faces.

- **aspergillosis**
a fungal disease of chicks resulting from inhalation of spores of asparagus from damp or mouldy utter of feeding-stuffs, causing the growth of white nodules in the lungs and resulting in breathily; difficulties.

- **asphyxia**
suffocation or the suspension of animation because of suffocation.

- **aspirating**
removing of chaff, dust, or other light material by use of air.

- **ass**
a small, usually grey, long-eared animal which belongs to the horse genus. Also known as donkey. It is commonly used as a draught animal. It is crossed with the horse to produce a mule.

- **assay**
to determine the purity or potency of a substance or the amount of any particular constituent of a mixture.

- **assimilation**
the utilisation and absorption of simple food substances.

- **associated compound**
any substance that is formed by the union of two or more molecules usually of the same or similar chemical composition to form a stogie complex molecule.

- **assertive mating**
a trend for non-random sexual reproduction in a population; tendency of certain types of males to breed with certain types of females, resuming in parents that are mentally and physically more similar than would be expected by chance.

- **aster**
the poles of a cell undergoing division. It is composed of microtubules, a clear zone and a pair of centrioles.

- **asthenia**
lack or loss of strength.

- **astringent**
drug that is able to contract tissues. It is locally applied to control bleeding relieve inflammation, etc.

- **astringent lotions**
liquids having one or metre astringent drugs. They often find use in the treatment of lameness in animals.

- **asymmetry**
the condition in which opposite sides of an animal are not alike; without symmetry.

- **asynapsis**
the failure of chromosomes to pair at meiosis or the absence. of chiasmata formation.

- **asystole**
the failure of the heart to contract, generally due to the walls having become so weak that they are unable to contract and expel the blood, with the result that hick organ becomes distended—a feature found after death.

- **atavism**
the reappearance of a character after a lapse of one or more generations, due to the character being recessive or being dependent on complementary factors, as a result of back insinuation; sometimes called throwing hack and the aberrant individual called throw hack. Also called reversion.

- **ataxia**
lack of normal coordination of parts of the body, especially inability to coordinate movement of voluntary muscles.

- **at-foot (suckling)**
thus an ewe with lambs at foot has suckling lambs.

- **atheroma**
a felly degeneration of the walls of arteries.

- **atherosclerosis**
1. a fatty degeneration of the connective tissue of the arterial walls.
2. a form of arteriosclerosis characterised by alheroma.
3. lesions within the arteries with plaques containing cholesterol and other lipid materials.

- **atom**
the particle of a matter indivisible by chemical means.

It is the fundamental building block of the chemical elements. The element such as iron and sulphur, differ from each other because (they contain different kinds of atoms. Atoms arc electrically neutral.

- **atmospheric air**
  the normal air that is around us all the time.
- **atomic number**
  the number of protons in the nucleus of an atom and also its positive charge.
- **atomic reactor**
  a nuclear reactor.
- **atomic weight**
  the mass of an atom with respect to other atoms- the atomic weight of any element is approximately equal to the total number of protons and neutrons in its nucleus.
- **atmospheric phase**
  the phase in a pulsator or line which is flooded with atmospheric air.
- **atmospheric pressure**
  the normal pressure of the air that is around us all the time.
- **atony**
  weakness of the system of any part of a body which is characterised by a lack of tone or lack of tension.
- **atopy**
  a clinical hypersensitivity state which is subject to-hereditary influences. This includes asthma, eczema and 'hay fever'. Canine atopy is characterised by sneezing, scratching, paw licking and face rubbing.
- **ATP**
  Adenosine Triphosphate, an energy-rich substance. It carries energy from biochemical reactions that yield energy to those reactions that use energy.
- **atresia**
  a defect which is existing in animals from birth, but it is seldom discovered until after the heifer gets freshened. Treatment is only surgical. To prevent the closure of leak orifice by healing a teat tube is inserted at milting lime and to replace it between milking with a sterile or dedicated teat dilator.
- **atrophic trinities**
  a chronic inflammation of the mucous membranes and

turbinate bones of the mice. It often causes distortion of the snout in twine. Growth rate is retarded. Secondary bacterial infection and pneumonia are common complications.

- **atrophy**

  a defect or failure of nutrition manifested as a wasting away or diminution in the size of a cell, tissue or organ.

- **attenuate**

  to make (micro-organisms or viruses) less virulent.

- **attenuation**

  1. weakening; reduction in virulence.

  2. any procedure in which the pathogenicity of a given organism is reduced or abolished.

- **attested area**

  an area in which a programme of specific animal, disease eradication is operating, declared as attested the disease having been reduced lo a very low level. Specifically applied to brucellosis and bovine tuberculosis atypical- irregular-abnormal— not conforming to type auditor of or pertaining to hearing or the organs of hearing.

- **auditory nerve**

  eighth cranial nerve of vertebrates going to the ear.

- **aujeszky's disease**

  a virus disease of cattle and pigs made noticed in 1979, it in characterised by intense itching of the hindquarters. Named after the Hungarian, aujessky, who first described it in 1902. Cattle lick the infected part bare and may bite it. Paralysis takes place within 24 hours and death within 4s hours. In pigs the disease is milder with less mortality characterised by loss of appetite, vomiting, diarrhoea convulsions, acute salivation and throat paralysis. Sows tiny about the disease also affect dogs and rats, which can transmit the disease by biting.

- **aureomycin**

  an antibiotic that has been passed to be effective against both gram-negative and gram-positive microorganisms and certain viruses, administered intravenously and orally. It is also used as antibiotic feed supplement.

- **auricle**
  chamber of the heart that receives blood from the veins.

- **auto catalysis**
  sometime a product of a reaction functions as a catalyst for it, the reaction is said to be auto-catalysed and such catalysis is called auto catalysis.

- **autoecism**
  capability of aparasitic fungus to complete its entire life cycle on the same host.

- **autecology**
  the ecology of an individual organism or taxonomic group, as opposed to synaecology which is the ecology of a community.

- **auiotetraplold**
  a tetraploid with four similar sets of chromosomes.

- **autogamous**
  self- fertilising.

- **autogenous**
  that is originated within the body of the patient.

- **auto immunisation**
  means re-infection from host's own parasites or the production in an organism of reactivity to its own tissues, with appearance of certain clinical and laboratory manifestations as a result of the altered immunological responses infection.

- **autointoxication**
  means self- poisoning which occurs due to the absorption of un-elifflinated toxins produced within the body, e.g., from waste products.

- **autolysis**
  breakdown of cells by the action of their own enzymes (endogenous enzymes).

- **digestive**
  the process of self-digestion, which lakes place in vegetable and animal tissues, particularly after they have ceased to be a normal part of the organism to which they belong; thus the tenderising of beef during banging, is an example of autotypic process.

- **automatic nervous system**
  self-governing, automatic nervous system consists of ganglia and motor nerves

to outstripped muscles anal glands which govern their involuntary action. It is divided into sympathetic and para-sympathetic system.

- **autonomic**
  1. self-controlling
  2. pertaining to that portion of the nervous system which regulates involuntary processes.

- **autopsy**
  examination, including dissection of a carcass to learn the cause and nature of a disease or cause of death.

- **auto radiograph**
  a photo graphic record of radiation from radioactive material in an object, made by placing the object very close to a photographic film or emulsion.

- **auto sexing breeds**
  breeds in which sex distinction is possible at day-old by means of down colour.

- **auto sex linked**
  pure breeds of poultry in which day-old male and female chicks respectively exhibit different colours.

- **autosomal genes**
  genes having their loci on chromosomes other than the sex chromosomes.

- **autosome**
  a normal chromosomet with its normal homologue i.e. a chromosome, other than a sex-chromosome.

- **autotroph**
  any micro-organism that uses only inorganic materials as a source of energy. Environmental carbon dioxide is the major or rather the sole source of carbon.

- **autumn fly**
  a non-biting fly that irritates cattle in large numbers and feeds on nose and eye secretions and open wounds. They cause cattle to huddle together and cease feeding.

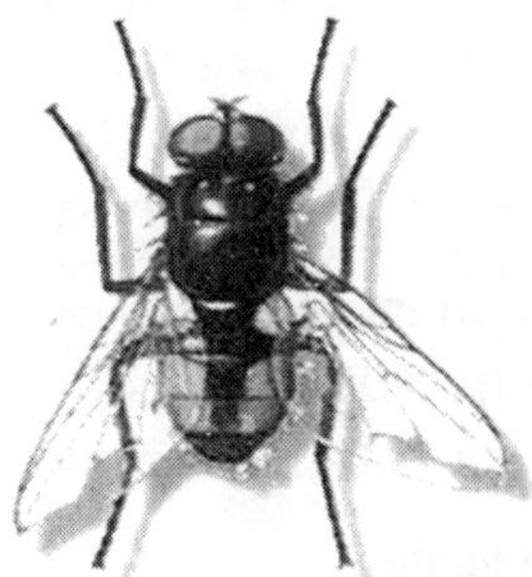

- **auxin**
  substance synthesised by the plant influencing growth of some point other than

the point of its synthesis, comparable to hormone in animals.

- **axial gradient**
  the gradation of the rate of metabolism.

- **auxotroph**
  strain of a micro-organism which requires some nutrient not required by the original strain from which it was derived. An auxotroph may arise by the occurrence of one or more mutations in a prototroph.

- **avian**
  pertaining to all species of birds, including domestic frills. A vain tubercle bacillus microorganism causing tuberculosis in swine and poultry but rarely in sheep or cattle.

- **avian tuberculin**
  used to diagnose the avian type of tuberculosis in farm animals particularly swain and poultry.

- **avidin**
  a specific protein component of egg albumen which encircles (combine with biotin to render ill unavailable to an animal. thus producing the syndrome known as egg white injury. the phenomenon is observed only when rats, for example, consume uncooked egg. because the avidin is denatured by heal and hereby iacliyaled.

- **avitaminosis**
  a term used to describe conditions produced by a deficiency or lack of a vitamin in the food.

- **axilla**
  the small hollow beneath the arm where joins the body at the shoulder (arm pit),

- **axis**
  a line about which a revolving body turns — a centre line.

- **axoti**
  the central core forming the semi-linial conducting ˙par of a nerve tire. The long extension of a nerve cell that carries impulses away from the body of the cell,

- **axoturia**
  a specific disease of horse found frequently in animals of draft type. It is chiefly occurred in the spring months

in animal that are receiving lull feed and are in good condition, but under-exercised. It is thought lo be caused due to excessive lactic acid, one of the by-products resulting from the

- **babcock test**
a test which is used for measuring the butterfat content of milk and cream, involving the use of sulphuric acid to break down the proteins which surround the fat globules in the reify, releasing the fat for measurement.

- **babial**
regarding lips.

- **babonic plague**
plague which is characterised by buboes (swelling of glands in the groin). Primarily rodents are affected by it. Later it is often transmitted to man and livestock.

- **bacillary white diarrhoea**
an often fatal bacterial of young chicks caused by salmonella pilfered, which causes serious looses between 4 and 14 days of age. It is inherited from parent stock which they have survived an outbreak carrier parent's can he detected by a simple blood test. Known as b.w.d. for short. Also called put.i.ohum disease.

- **bacillus**
a genus of spore- producing bacteria. Also a general term for rod shaped bacteria.

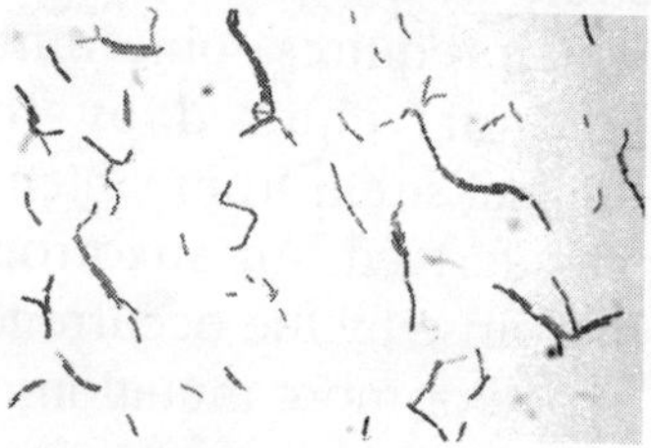

- **bacitracin**
a complex antibiotically active chemical compound belonging to the polypetide class produced during the fermentation of suitable culture media by a special strain of the bacterium bacillus subtilis.

- **back chain**
a chain fastened across the back of a draught horse to the ridge roils on the shafts of a cart or wagon.

- **backbone**
the vertebral column, especially if composed of bone.

- **backcross**
the crossing of a hybrid with one of the parcel types

(breeds). The offspring are referred to as the backcross generation, or progeny.

- **background count**
  the number of impulses per the time registered on a counting instrument when no sample is present.
- **backwasher**
  a machine for washing wool after it has bee a carded.
- **bacon**
  the meat from the back and sides of a pig, preserved by curing with brine. Bacon may be sold 'green' or after wood-smoking which produces strong-flavoured 'smoked bacons'.
- **bacon side**
  a half carcase of a bacon pig.
- **bacteria**
  large group of one-celled, non-spore forming micro organisms, which may appear singly or in colonies as spherical, rod-shaped, or spiral threadlike cells.
- **bacterial count**
  the number of bacteria, just viable or both viable and non-visible depending on method of analysis.
- **bactericidal**
  quality of an agent that is capable of killing a bacteria. It can be any physical or chemical agent that is capable of killing some types of bacteria.
- **bactericide**
  an agent or substance capable of destroying bacteria.
- **bactericidin**
  substance present in the sera of mammals that is capable of inhibiting the grown up bacteria, especially the gram positive ones, or of destroying them.
- **bacterin**
  vaccine containing attenuated or killed bacteria administered to increase the resistance of an animal to a specific infectious disease.
- **bacteriophage**
  an ultramicroscopic, filterable virus composed of deokyribonucleic acid and a protein coat. Infects particular strains or species of bacteria and usually causes lyses, or species of bacteria and usually causes lyses, or explosive dissolution of she infected bacterium. Bacteria pages multiply at the ex-

pense of bacterial cells. They have no enzymes of their own.

- **bail milking**
  the milking of cows inside a portable shed or bail, either moved from he'd to field frequently or kept in a relatively permanent and sheltered position. Sometimes the cattle are kept in temporary or permanent yards during the winter and milked in an adjacent bail bailiff. A person employed to look after or manage a farm on behalf of the owner, receiving either a salary or share or the profits.

- **bait**
  a mixture of oats chaff, hay and other dry food used as horse feed.

- **balanced**
  containing essential nutrients in proportions.

- **balanced daily ration**
  a combination of feeds to provide the essential nutrients in such amounts as will properly nourish a given animal for a 24-hour period.

- **balanced feed**
  a feed laving all the known required: nutrients in proper amount and proportion based upon recommendations of recognised authorities in the field of animal nutrition. The species of which it is intended and conditions of growth and development should be specified.

- **balanced ration**
  the ration, which contains a combination of feed ingredients that provide all essential nutrients in proper amount and proportion based upon the standard recommendations for a particular species under a specified physiological condition such as during growth, pregnancy or in production where the quality and quantity of the products are also considered.

- **bale sledge**
  a sledge towed behind the delivery point of a ram baler, on which the operator stacks bales (usually 6 or 8), although many are now unmanned. these stacks are

removed from the sledge periodically.

- **ball**

a bolus for a horse.

- **balling guns**

metal instruments used for the administration of medicated capsules or large tablets to sheep, swine and horses.

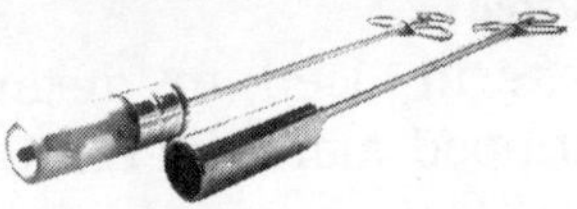

- **ballottement**

a technique of clinical examination in which the movement of any body or organ, suspended in a fluid, is detected.

- **band seeding**

placing forage crop seed in rows directly above, but not in contact with, a band of fertiliser.

- **bandage**

a piece of doth or elastic material applied lo retain surgical dressings or prevent motion, etc.

- **bantams**

diminutive forms of large poultry.

- **bar pad**

a leather sole for a horse's foot which carries a rubber or leather pad across the whole of its posterior portion and is fitted along with a short shoe bar pig.

- **barb**

1. a dark-coloured fancy pigeon
2. a swift breed of horse.

- **barbon**

a farm yard.

- **barley feed**

the by-product obtained during the manufacture of pearl barley from clean barley.

- **barley hulls**

the outer covering of barley.

- **barley mixed feed**

the products obtained during the milling of barley flour from clean barley. The compound is mostly composed of the outer covering of the barley (hulls) with coarsely ground barley particulars.

- **barn**

1. a unit of area for nuclear cross section.
2. refers to shelter constructed for cattle, horses and forage crop

- **barn allergy**

an allergy, similar to hay fever, the symptoms of which include a runny nose, sneezing, inflammation of the eyes or more severe symptoms. Bam allergy symptoms occur at the time of contact with dust from hay or grain. The principal cause is the presence in stored feeding-stuffs. The most effective solution to the allergy is the use of a respirator.

- **barn cured**

cured within a barn, as contrasted to sun or air-cured.

- **barn machinery**

implements or machines that are not used in the fields, e.g. Hammer mill, grain drier, weighing machines, chaff cutters, etc.

Bentall Green Bone Cutter BCB3. Green Bone Cutter BCD3.

- **barnevelder**

a heavy breed of fowl, with black and white varieties, although similar in colour to a partridge. It has a comb; yellow legs and the hens produce dark brown barnyard, land immediately adjacent 10 bran, normally fenced in to enclose livestock or fowls.

- **barred**

a term used to describe stripped markings on fowl.

- **barred feathers**

poultry feathers exhibiting a horizontal pattern of black and white stripes or bats. Other colours may be combined with black such as buff or gold, thus giving buff-barred and gold- barred feathers,

- **barren**

incapable of producing offspring, due to infertility or sterility.

- **barrier**

any type of obstruction- biological, chemical or physical that does not allow the migration of animals, or gradual extension if their territories.

**barrow**

a male hog whose testicles were removed before reaching breeding age and before the development of secondary sex characteristics.

**basil**

the finished dressed leather obtained from sheep skin.

**basket brooder**

brooder made with an ordinary dome-shaped basket, lower half of which is fined with gunny sack or cement mixture wash and upper half left open to allow passage of smoke hurricane lantern sources of heat, sleeved off by 1-2 inch wire mesh: door covered with straw or raw dust.

**batch drying**

a method of frying the bales of hay or batches in a barn, usually using seated air blown through a ventilated floor. Batches of about four layers have been dried, at a lime until moisture content get reduced below 20, are then removed fur storage and the next batch introduced this method is less popular nowadays than storage drying and tunnel drying.

**Bateson, William**

british geneticist (1861-1926) who influenced the science of genetics by studying genetic variation as a basis for evolutionary change.

**bating**

one of the steps used in the tanning of leather. It follows de-hairing. In this the skin are soaked in a bath containing chymes and ammonium sails.

**battery hens**

hens kept indoors in cages for laying, normally in three-tiered rows, laying bactenes may be static, semi or fully-automatic, depending on the method of feeding, watering and manure clearance. Battery housed birds achieve a higher rate of egg production than those under other systems.

**bay**

1. the combined cry of hounds hunting an animal, or the last stand of an animal cornered by hounds,
2. a description of a horse, reddish brown to chestnut in colour.
3. the space between two columns in a bam, or any recess in a farm building.

4. a stall in a stable.

- **beam hole**
an opening through a reactor shield which permits a beam (stream) of radioactive particles or electromagnetic radiation to be used for experiments outside the reactor.

- **beamage**
an animal carcase or an allowance for loss of weight by evaporation.

- **bean**
the bean-shaped tip at the end of a duck's bill.

- **beard**
1. a bunch of feathers beneath the throat in certain breeds.
2. the hairy tuft on the lower part of a goat's jaw, or on a turkey's neck

- **bedding**
litter for farm animals to sleep on, usually straw.

- **beef**
meet which is derived from cattle of nearly one year old or older.

- **beef bull**
a bull which is reared and fattened to produce beef. Male cattle for fattening are usually extra red to produce steers. Bull carcases are leaner than those of steers, with darker muscle and a heavier forequarter. The advantage of beef bull production is that bulls grow faster, convert food more efficiently and achieve greater carcase weigh than steers,

- **beef cows**
heifers and cows which are kept mainly for rearing calves for beef production, as distinct from dairy

- **beef shorthorn**
a breed of compact, short-legged beef cattle.

- **beefy**
the desirable physical conformation of a beef animal as contracted with a dairy animal, which is lean (not beefy) and more angular.

- **beet molasses**
beet is also utilised for the manufacture of sugar apart

from cane sugar, the product is known as beet sugar. During this process a by-product is obtained known as beet molasses, carries the same specifications as cane molasses.

- **beet pulp**
  a fibrous by-product of beet sugar industry, bees sugar.

- **behaviour**
  the reactions of the whole organism to its environment.

- **belgian blue**
  a breed of cattle found in Belgium, South Netherlands and Luxembourg.

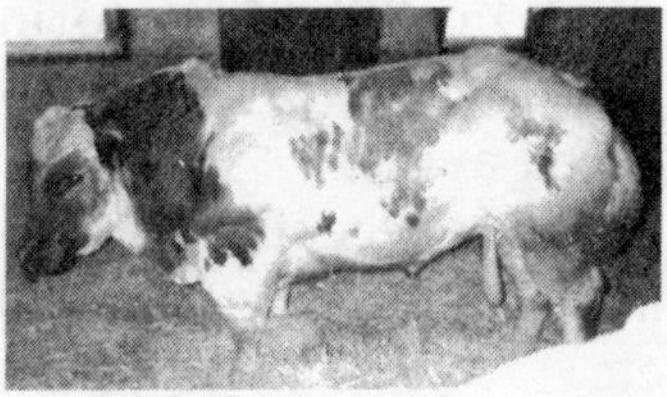

- **bell wether**
  the leading sheep of a flock, with a bell on its neck.

- **bellies**
  the wool from the belly or stomach area of the sheep.

- **belt**
  1. a term for sheep scab
  2. a zone of country

- **benign**
  1. mild or not serious
  2. not malignant.

- **beta particle**
  an elementary particle (an electron) emitted from a nucleus during radioactive decay. Beta radiation may cause skin burns, but a thin sheet of metal easily stops beta particles.

- **bighead**
  an animal disease affecting ewes, lambs and goats during summer. It is characterised by swelling of the parts of the head that are bare or lightly covered with wool. Seriously affected cases require more care. Bathing the affected parts with an epsoro talt solution is also effective. The disease could be prevented by confining the flock in a shed, barn or some other shaded area during mid-day.

- **bile**
  a yellowish or greenish brown fluid secreted the liver; it flows through the liver's bile ducts into the first portion of the small intestine. Bile consists of sodium salts of bile acids, which help, in emulsifying fats and

a number of bile pigments.

- **bile salts**
these are alkaline compounds formed in the liver and discharged into the small intestine as part of the bile they consist of the sodium salt of chilli; acid combined with turbine or with glucose and assist in the emulsification of fats by lowering the surface tension of fat droplets.

- **biliary pigments**
pigments formed in the liver from the haemoglobin of old red cells which are destroyed there.

- **big leg**
an animal disease that develops in the front or hind legs and may extend to the shoulder, neck and rum. It is possibly caused by vitamin A deficiency. The affected animal loses weight rapidly and moves about slowly. Occasionally saliva drools from the mouth. Mild cases respond in 3 to 4 days when the animals are turned on pasture and fed new corn.

- **biliary canals**
the network of bile ducts in and around the liver cells.

- **billy goat**
a male goat, also known as a buck.

- **bilocular**
having two chambers.

- **bimodal distribution**
a frequency distribution with two modes.

- **bin**
1. a canviis container on a collapsible wooden frame used initially for storing hops when picked. Also called a crib.
2. a large receptacle for storing feeding stuffs, corn, fruit, etc.

- **binders**
fibres that hold the wool, staples together, enabling the fleece lo be handled as a unit.

- **bioassay**
the use of animals to determine the active power of a compound as compared with the effect of a standard preparation.

- **biochemistry**
  the chemistry of living things.

- **bioclimatology**
  the field that studies the effect of physical environment upon living organisms.

- **biocoenotic**
  the inter-relationship between organisms in a bioengineering, the manufacture and use of artificial replacements for body organs that have been lost of are malfunctioning, such as artificial limbs, heart pacemakers.

- **biogenesis**
  the doctrine that living organisms arise only from other living organisms and never from non-living matter, it also slates that organisms can only arise from others like themselves, i.e., of the same species and not from very different unrelated ones.

- **biogenetic law**
  also known as the recapitulation theory, it is the principle that an animal may repeal in its embryonic development some of the corresponding stages of its ancestors.

- **biolimatographs**
  graphs in which flora or fauna distribution and climatic factors are plotted simultaneously to show the climatic factors which limit the distribution of a given species.

- **biological**
  pertaining to the science of life.

- **biological control**
  the destruction or suppression of un-desirable insects, other animals, or plants by the introduction or propagation, encouragement, artificial increase and dissemination of their natural enemies, which include predaceous and parasitic insects, predatory vertebrates, nemaiode parasites, bacteria, protozoa, viruses and parasitic fungi.

- **biological function**
  the rule played by a chemical compound in living organisms.

- **biological half life**
  the time required for a biological system, such as a

man or an animal to eliminate by natural process half the amount of a substance (such as a radioactive material) that has entered it.

- **biological mineralisation**
  the conversion of an element occurring in organic compounds to the inorganic form as a result of biological decomposition.
- **biological value of proteins (bv)**
  the percentage of the absorbed protein which is utilised by the body. In this case losses in digestion and metabolism are taken into consideration.
- **biological year**
  the time interval between the first and the last egg laid before a hen enters in moult.
- **biologicals**
  medicinal preparations made from living organisms and their products, used for the prevention or detection of disease: they include serums, vaccines, antigens and antitoxins.
- **biomass**
  the total weight of living matter in a population, it is usually expressed in terms of dry weight per unit area.
- **biometry**
  study of variation in biological organisms.
- **bionomics**
  the study of the relations of organisms to their ecology and environment.
- **biopsy**
  the chemical analysis or microscopic of issue removed from a living body.
- **biosynthesis**
  bringing together of chemical building units to form new materials in the living plant or animal.
- **biotic**
  pertains to life or living matter; biological.
- **biotic factors**
  ecological influences that arise from activities of living organisms as distinct form e.g., climatic factors.
- **biotic potential**
  an estimate of the maximum rate of increase of species of animal if left to itself and isolated from biting factors. Normally this potential reproductive rate is very great but in the struggle for existence a

balance is maintained and the population of any species remains roughly consistent. Sometimes, however, conditions such as an extra long warm summer may enable certain insects to product an additional generation in the same season. This results in a 'population explosion' or plague of some particular insect. Usually, however, predators that feed on these insects will increase in proportion and the balance is soon restored.

- **biotin**
  one of the vitamins.

- **bird egg**
  a very large reproductive cell hatched by birds.

- **birtch**
  the thigh and twist part of a sheep and also the wool from it which is coarse and of low quality.

- **bitch**
  a female dog.

- **bite grazing**
  early bite is grazing in the early spring. Late bite is grazing at the end of growing season.

- **black blowfly**
  it develops in decaying meat.

- **black disease**
  an acute malady a fleeting mature sheep in good conditions. The germ ciosiridiurti nautili in the presence of liver flukes cause it, it occurs on swampy, poorly drained land in the late summer. The first symptom is usually the finding of dead sheep in the morning although the height before everything seemed to be in order. The dead animals appear to have died in their sleep.

- **black faced sheep**
  the name used by western ranchmen to designate sheep of the down breeds or crosses of these breeds with other sheep. In Great Britain it is the name of a scotch sheep that produces very

coarse wool.

- **black vomit**
it is due to the presence of blood in the stomach. The appearance of the vomit may be either that of black masses of clothed blood, or it may resemble coffee-grounds.

- **blackhead**
a disease of turkey caused by a protozoan, nteleagridi.1, the liver amid intestines. Symptom includes ruffling of feathers, hiss of appetite and mustard-yellow diarrhoea.

- **blackleg**
an acute, infectious disease of cattle which is characterised by swelling beneath the skin, due to gas formation in the tissues. It is caused by the bacterium clostridium chauvei. The infection spreads through food, water and wounds. Penicillin has been found effective in treating affected animals. Immunisation by means of vaccination is the most practicable and effective means of protecting individual animal

- **blackleg bacterin**
cultures of the blackleg organism suspended in the medium in which it was grown. It is used as a vaccine against blackleg disease of cattle, sheep and goat.

- **blacks**
black fleeces, as reported to the wool growers by dealers or cooperatives who sell wool on commission. Black wool generally sells at a lower price than corresponding while wool.

- **blastema**
a group of cells that give rise to an organ or structure either during regeneration or in normal embryo genesis.

- **blastocyst**
the name given in a very early slog in the development of the foetus, in which

the mass of cells, resulting from the fertilisation of the ovum, which is to become the new animal, assumes a hollow, spherical at first and hen ellipsoidal, bladder like form,

- **blastula**

stage in embryonic development in which cells are arranged to form a single-layered hollow sphere.

- **blaze**

a white mark on an animal's face.

- **bleat**

call of sheep and goat

- **bleeder**

an animal that has haemophilia.

- **blend**

1. to cross-breed plants or animals to obtain an off-spring having characteristics of each,

2. to mix two or more ingredients to get a standardised product.

- **blending**

1. for marketing— the physical mixing in small lines differing slightly in type or yield to produce large even lots when repacked for sale,

2. to manufacture—the mixing of wools of different type or origin to produce a special effect, a product with constant characteristics, or the reduce costs.

- **blending inheritance**

the inheritance of characters so that the off-spring and successive generations become intermediate between the original parents. This is because that character has been controlled by several genes.

- **blepharitis**

inflammation of the eyelids. It is usually associated with conjunctivitis.

- **blind quarter**

a quarter of an udder which does not secrete milk or one that has an obstruction in the teal which prevents the removal of milk. A non-functional mammary gland.

- **blind teat**

a condition in which the end

of the teat has failed to emerge.

- **blistering ointments**

are employed as irritants and counter-irritants as in the treatment of lameness they are often used to support the effect of firing.

- **blisters**

these are caused due lo irritation of the deeper skin layer which causes a fluid lo accumulate. Pus-filled blisters are called pustules.

- **bloat**

an abnormal accumulation of gases chiefly hydrogen sulphide and carbon monoxide in the rumen of livestock, resulting in great distension of this organ with consequent paralysis of its waifs. It is caused by vigorous fermentation in the rumen due to eating too fast and too great a quantities of legumes. Controlled feeding could prevent it. Treatment of affected animal consist in passing a stomach tube through mouth and throat; stimulating breaching by placing a wooden bit 5 to s cm diameter in the mouth like a bridle; making animal stand with front feel as high as possible and administrating linseed oil or white mineral oil.

- **blocky**

term commonly applied to meat-producing animals.

- **blood**

1. the fluid tissue that fills the blood vascular system. It consists of a liquid containing blood cells, platelets, dissolved respiratory gases, digestive and excretory products, proteins, hormones, salts etc. It often contains respiratory pigment, either free in the plasma or in cells.

2. the terms half blood, three-eighth blood, quarter blood and low quarter blood are the names of four grades of won . As used at present, they are arbitrary names to indicate degrees of fineness of wool and not the breeding of the sheep from which it came.

- **blood agar**

a highly nutritive medium containing red blood cells. It promotes growth and production of haemolysins by most micro-organisms, especially pathogenic bacteria.

- **blood cells**
cells suspended in blood serum, e.g. RBC, WBC, haemoglobin.

- **blood diarrhoea**
acidosis of ruminants swine, poultry etc.

- **blood feedstuffs**
dried blood, e.g., blood flour, blood meal etc.

- **blood groups**
types of blood classified on the basis of the different antigens present on the surface of the red blood cells, which are genetically determined. The antigens are distinguished by means of specific serum. There are four groups: A, B, AB and O characterised by the presence of antigen a, antigen b, both antigens A and B or no antigen, respectively on the red blood cells. A person of blood group a can receive a blood transfusion only from other people of group A or from group AB.

- **blood meal**
1. a residue from a slaughterhouse, rich in protein, iow in mineral content, used as a feeding stuff particularly in rations for pigs and poultry.
2. blood which has been dried firstly by passing live steam until the temperature reaches 100°c, the process ensures sterilisation and causes blood to clot. It is then drained, pressed to expel occluded serum, dried again by steam heating and finally ground. The compound contains about 80% protein but deficient in leucine.

- **blood plasma**
the liquid portion of blood in which the corpuscles of blood cells are suspended.

- **blood poisoning**
it develops in form of scepticism or pyemia in many diseases of livestock, especially in anthrax, shipping fever, abortion, mastitis etc. It is the cause of death in the very acute form of calf diphtheria. Blood poisoning is often due to wound infection.

- **blood pressure**
the pressure of blond in the main arteries of mammas. In

resting humans, blood pressure normally oscillates between about 120 mmHg at systole about 80 mmHg at distole. Blood pleasure varies in response to climatic conditions, exercise and posture.

- **blood serum albumin**
chief water-soluble protein in the serum of blood.

- **blood sports**
those sports involving the killing of animals. e.g., foxhunting.

- **blood stock**
thorough-bred horses, bred through generations for their excellent qualities.

- **blood transfusion**
the transfer of blood from one animal to another. This is of great importance in the treatment of some disease like equine influenza, wrangles, and haemorrhagica. Blood may be transferred directly from a vessel of the donor to that of the 'receptor' by means of a connecting tube. Indirectly, the donor's blood can also be collected in a vessel and later used as citrated blood or as serum.

- **blood vessel disorder**
normal tone and function of the body muscles are dependent upon the supply of blond they receive. Inflammation of the bleed vessels, chiefly the arteries,
May occur in the legs as well as other parts of the horse body as a result of invasion by strongyles. The walls become thickened and clots form in the iliac, femoral and other arteries of the leg. The blood la the legs are decreased. Clot formation in the iliac artery produces a characteristic type of lameness with which there may be profuse sweating, trembling and rapid breaching. The attack may disappear as rapidly as is developed, leading the animal entirely normal, or a part of the dot may be dislodged into the blood stream and become fixed in a vessel in some vital organ, sued as the brain, which may cause death.

- **bloodline**
succeeding generations of animals of a species which are specially bred to maintain the genetic composition responsible for specific

desirable features or qualities. Maintaining a bloodline is also known as *breeding true*.

- **bloom**
  a term commonly used to describe the beauty and freshness of a cow in early lactation, a diary cow in bloom has a smooth hair coat and presents evidence of milking ability (dairy character). Also, a term expressing strength of gel in gelatine.

- **blow**
  the condition of a cow with its stomach swollen by gases, mainly methane,

- **blow fly**
  1. the bluebottle, a two-winged fly with metallic blue winger the larvae feed on animal matter and dung.
  2. a general name for a number of species of fly which deposit their eggs in flesh and carcases. Also called flesh fly.

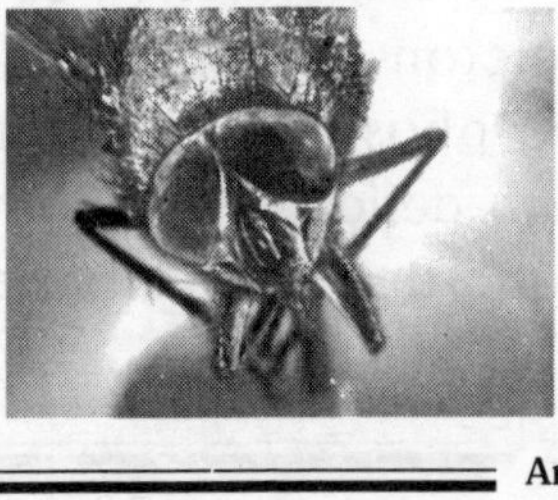

- **blowing**
  the process of blowing dry steam through a cloth, to settle the fabric and take the curliness from the yarn.

- **blown**
  1. a descriptive term for an animal suffering from bloat.
  2. a descriptive term for sheep suffering form attack by maggots, particularly of the blow fly.

- **blue ointment**
  mild mercurial ointment having a bluish colour. ft is used for the destruction of external parasites, particularly for the extermination of lice.

- **boar**
  a male hog with sex organs intact and generally used for breeding purposes. (the uncastrated male of swine).

- **bobby calf**
  the unwanted male offspring of a milk- producing breed of cattle, usually a Channel Island breed, used for veal.

- **body**
  wool when the staple appears full and bulky.

- **bog spavin**
  the accumulation of synovial fluid in and around the hock joints of horses.

- **boiler**
  a pullet or hen which is sold for the table at the end of the laying period.

- **boiling point**
  the temperature at which the vapour pressure of a liquid equals the atmospheric pressure.

- **bold**
  a term applied to well-grown wool of good character.

- **bolus**
  regurgitated food that has been hewed and is ready to be swallowed as a large pill for dosing animals.

- **bomb calorimeter**
  an apparatus for measuring the heat of combustion, as of human foods, animal feeds and fuel.

- **bone**
  the material of the skeleton. it consists of fibrous tissue, which forms gelatin if boiled,

- **bone ash**
  is obtained by burning bones with free access to air and contains 15 to lfi.5 per cent phosphorus. i.e. used as a feed supplement.

- **bone charcoal**
  bone black is the product obtained by charring (partial burning) of bones in closed retorts (a container generally of glass with a long tube, in which substances are decomposed by heat).

- **bone chewing**
  animals on feed which are low in phosphorus commonly chew bones, wood dried or decayed carcasses as an instinctive reaction to the dietary deficiency affected animals get unthrifty and milk production reduces 10 a minimum. The remedy involves feeding of phosphorous rich feeds such as oil cakes.

- **bone disease**
  various abnormal bone conditions are characterised by rather definite alteration in the conformation of the skeleton. Several factors like phosphorus and calcium deficiency, abnormal proportion of calcium and

phosphorus in the ration, deficiency of vitamin D and A and an altered rate of secretion of certain endocrine glands may be involved. Remedy involves removal of cause by suitable correction in the diet. Often it is necessary to correct protein deficiency, which is coexistent with a low phosphorus intake.

- **bone feedstuff**
  a common supplementary source of phosphor for animals. It is bones in variousformx provide a natural source of our calcium phosphate which they supply phosphorus and calcium in animal diet.

- **bone flour**
  finely ground bones containing phosphate, used as fertiliser and also as a feeding stuff.

- **bone meal (steamed)**
  a mash of ground animal hones that were previously steamed with pressure. It contains 1.5 to 2.5 per cent phosphorus and 20 to 34 per cent calcium. It is used as a fertiliser and as feed for farm annuals.

- **bone seeker**
  any compound or ion that migrates in vivo-preferentially into bone.

- **boning**
  taking out the bones of a fowl for table use, bonnet- the second stomach of a ruminant.

- **bonney clabber**
  an Irish term for milk naturally clotted on souring.

- **boost**
  an instrument used to mark sheep, often with the owner's initials, usually using hot tar.

- **booted**
  fowls which have feathers on the toes, for example, Brahma.

- **boots**
feathers projecting from toes as in the brahma, etc.

- **boric acid**
white powder used in medicine as very weak antiseptic.

- **bottle jaw**
an edema, which develops under the jaw.

- **bottom growth**
those grasses and clovers in a pasture growing close to the ground, as opposed to the taller plants.

- **botulism**
food poisoning which occurs due to the toxin of ctostridium botulinum. Fungal food poisoning lakes place in man and animals. Fungal toxins include, aflatoxins.

- **bound stock**
those animals, which usually have been bred on a farm, such as a flock of sheep or a dairy herd, which remain with the farm when it is sold or changes hands. also used to mean sheep acclimatised to a Hirsel.

- **bovine**
pertaining to the ox or cow.

- **bovine growth hormone**
a protein substance produced and released by the pituitary gland which is located at the base of the brain.

- **bovine ketosis**
a disease of cattle caused by an excessive drain on blood sugar for the production of lactose, which impairs energy metabolism and results in accumulation of toxic levels of ketones in the blood, milk and urine. Symptoms include loss of appetite, sweet-smelling breath to milk (due to the presence of acetone and paratactic acid), falling milk yields, loss of weight, lethargy and nervous behaviour. Most animal recover after treatment with glucose, glucose-forming substances or gliconeogenetic hormones.

- **bovine leucosis**
this is a leukaemia like disease in cattle. It may remain benign throughout life, or become malignant usually in cows 4 to 8 years old, resulting in fatality.

- **bovine malignant catarrh**
highly infectious disease which is characterised by

fever, catarrhal inflammation of the oral, nasopharyngeal and conjuclival mucous membranes, general lymprioid hyperplasia and frequent involvement of the central nervous system.

- **bovine tubercle bacillus**
  it causes tuberculosis in cattle, swine and goats.

- **bovine tuberculosis**
  a chronic contagious bacterial disease caused by mycobacterim tuberculosis.

- **bovicola**
  blood sucking lices.

- **boxed**
  when different mobs or flocks of sheep are mixed.

- **branding**
  1. of sheep — the placing of a registered brand on the sheep with u branding fluid for purposes of identification. Firebrands are sometimes placed on the face or horns of sheep.
  2. of wool bales— stencilling to signify he owner, serial number and type of wool in the bale.

- **break**
  1. a temporary interference with the growth of the wool, causing a marked thinning of all or a proportion of the fibre population and producing distinct weakness is one part of the staple. It is caused by a sudden change of pasture, want to food or water, sickness, bad lambing, or faulty dipping
  2. to cut up an animal's body.
  3. to tame or train a horse to wear a saddle or for draught work,
  4. a change of crop in a rotation programme.

- **break through**
  that point in an absorption cycle at which the effluent begins to show a marked increase in concentration of the substance being adsorbed.

- **breakaway**
  a stampeding animal.

- **breast**
  the front of the body from the pot of the breast bone up to the throat, in a live fowl; in a dead table fowl the breast in each side of the breast bone,

- **breech**
  the coarsest woof, which

comes from the hindmost part of the fleece. it is often hair-like fabrica,

- **breeching**
  a strong teacher strap on a harness passed round a horde's haunches, attached up the saddle und the shafts of a cart, allowing the horse to reverse the cart.

- **breed**
  1. a group of animals descended from common ancestry and possessing certain inherited characteristics which distinguish it from any other group.
  2. animals who are genetically pure enough so as to have similar characteristics of colour and conformation and when mated together will produce offspring with the same characteristics.

- **breed average**
  the average milk production of cows for a given diary breed. (Usually, computations arts based on the DHIA records of the past five years).

- **breed average herdmates**
  daughters of a sire in the same herd producing at the current average production for that particular dairy breed.

- **breed character**
  the physical characteristics that distinguish one breed of animals from another within the same species.

- **breed count**
  a microscopic method of counting bacteria in a dried, brained film of milk.

- **breed in, breed out**
  to introduce or remove a characteristic from an animal breed or plant variety by continuous breeding of those individuals lacking or having the characteristic, respectively, until it become lined in or is fit for the breed or variety (in breeding, character).

- **breed index**
  term applied lo the average number of services per conception in a dairy herd, gives indication for checking the bull and the cows when rate is higher than normal.

- **breed true**
  to have the ability to transmit a characteristic uniformly to offspring.

- **breed type**
  a particular type or form characterising a breed. It includes special breed features in head, ear, colour, or other traits common to a particular breed.

- **breeding**
  the genetic improvement of plants or animals by selection, hybridisation etc.

- **breeding chute**
  a wooden enclosure which holds a mare or cow or she-buffallo to facilitate mating.

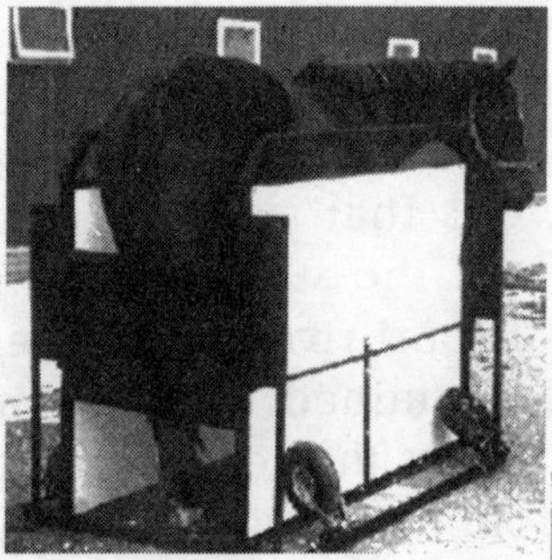

- **breeding crate**
  ad apparatus designed to take the weight of bull which is too heavy for the heifer or cow with which she is to be mated, also called a service pen or service crate.

- **breeding rack**
  equipment used to restrain females so breeding can be accomplished with little danger to animals or humane

- **breeding stock**
  farm animals selected for producing offspring as opposed to being fattened for slaughter, etc., in order to maintain or increase the size and quality of a herd of flock.

- **breeding trouble (animal)**
  it includes any cause of barrenness in either the male or female and the abnormal termination of pregnancy in abortion and stillbirth and retained after birth. The method of handling breeding animals has a decided influence on their physical conditions and health. An upset in the secretion of any of the sex. In the female it may interrupt the co-ordinated sexual rhythm, prevent conception, or induce early or late abortion.

- **breeding value (genetic value)**
  the genetic ability or' an animal to secrete milk, lay eggs, etc. one half of this genetic ability is transmitted to soils or daughters.

- **brewer's dried grain**
  the grain product, are the

residue which remains after most of the starch and sugars have, been removed from the barley malt alone or in combination with other cereal grain products in beer industry. The compound has got about 25% crude proteins. In view of high fibre content the product is seldom fed to swine or poultry.

- **brewer's dried yeast**

dried, non-fermentative non-extracted yeast belonging to the species saccharqmyces cerevisiae obtained as a by-product in the process of brewing beer, ii contains about 42% crude protein. The compound is a valuable source of many of the b group of vitamins and also relatively rich in phosphorus.

- **bricky**

a dull harsh red colour, usually found on buff cockerels and dark topped black-red game cocks.

- **bridle**

an apparatus fitted to a horse's head to control it, including the bit and reins.

- **bright wool**

the wool that is whitest and most attractive in colour which comes largely from the northern states east of the Mississippi river, as contrasted to the semi -bright wool that comes largely from the states along the Missouri river, or the deny or stained wool that reaches the central markets from time to time from various places.

- **brightness**

a term used to describe the white colour and light- reflecting power associated with the fine types of wool.

- **brim**

1. to put a boar to a sow for mating.
2. a sow on heat.

- **brisket**
  the breast of an animal. The term is used for the cut of meat from next to the ribs.

- **britch or breech**
  the back portion of the sheep down the hind legs; the buttocks.

- **british breeds**
  those native of Great Britain, such as Hereford, agues and shorthorn.

- **brittle hoofs**
  an abnormally dry slate of the horn. The hoofs become almost of the consistency of stone, chip and crack easily, cause-contracted heels and lead to difficulties in shoeing. Long-continued dryness or stabling on dry hard floors it conducive to the trouble. It is advisable to remove the shoes, trim feet and reshow. The bearing surface should be made levelled using hoof cement if needed. Hoof ointments are also beneficial.

- **broad**
  a term applied to woofs which art on the strong side- for their quality number or for their type.

- **broad spectrum antibiotic**
  an antibiotic that is active against a large number of microbial species,

- **broken**
  a trade term applied to the best wool of the skirting, having the characteristics of fleece wool.

- **broken horse**
  a horse trained to the saddle or bridle.

- **broken mouth**
  a sheep is described as 'broken mouthed' when some of the incisor teeth have fallen out or become badly worn and irregular, usually as a result of old age or hard grazing.

- **broken udder**
  a term sometimes used to indicate an udder that is pendulous and loosely attached.

- **broken winded**
  having short breath or defective respiratory organs and thus incapable of hard work. Particularly applied to horses.

- **brome grass**
  a large grasses strongly resembling oats, but more

flowery, mostly unimportant at fodder and generally regarded as weeds in Britain and becoming increasingly troublesome.

- **bronchiole**
  one of the many, very fine sub- divisions of the bronchial tubes.

- **bronchitis**
  refers to the inflammation of the mucous membrane of the bronchial tubes.

- **bronchopneumonia**
  the inflammation of the bronchi and of the lungs.

- **brooder**
  also called *rearer, brooder house and foster mother*. Important equipment used in artificial brooding of chicks for temperature and regular supply of fresh air providing dryness and easy disinfections, protection against enemies, made in various designs, depending upon nature of fuel supply and size and number of chicks to be reared,

- **bloodiness**
  desire of birds to sit in a net on eggs for the purpose of hatching. It is the maternal behaviour for hatching and rearing.

- **brooding**
  the rearing of chicks after hatching till maturity; natural and artificial methods followed.

- **brooding coop**
  coop or cage which has been provided to hen and the chicken to protect them from enemies and inclement weather, essential requirements of which being dryness, roominess and safety; bamboo coops, barrels, packing boxes and a-shaped coops commonly used.

- **brooding stage**
  the period, 4-8 weeks, from hatching until poultry chicks cease in need some form of heal for survival, provided by a broody hen under the protection of its wings, or artificially in 3 brooder,

- **broody hen**
  a hen inclined lo sit on its eggs to hatch them, also called a sister.

- **brown dog tick**
  a pest of dogs which is sometimes found on livestock.

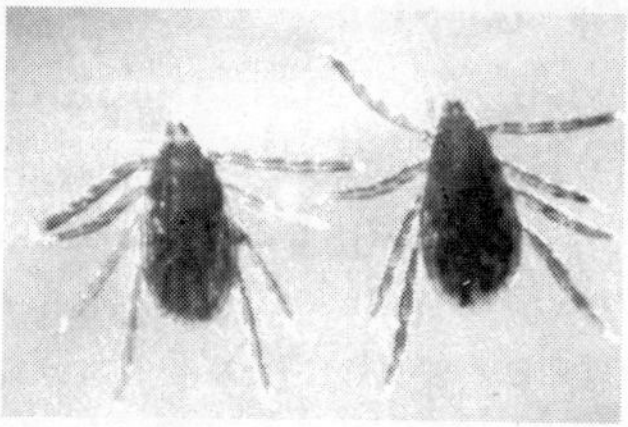

**brown heart**
a deficiency disease of swine due to lack of boron, causing browning or mottling of the toot making it unpalatable.

**brown nose**
a form of light sensitisation in cattle.

**browse**
palpable portions of woody vegetation grazed by brucella abortus vaccine, it is obtained from a strain of brueeita aborfut whose virulence has been reduced. It is used for the vaccination of calves between 4 and 5 months of age and is harshly effective in preventing brucellosis.

**brucellosis**
a bacterial disease of the reproductive system, mainly to cattle, sheep and pigs, although all mammals have been susceptible. In cattle it causes infection of the womb and plasma and abortion of calves. It is spread mainly via aborted calves, afterbirth, discharges and milk, which contaminate pasture, food and water. It causes undulant fever in humans. Also known as contagion

**bruise**
a tissue injury without laceration. The characteristic colour of a bruise is due to blood, which has been poured out into the tissues underneath the skin.

**brushed or brush pulled**
in modern wool pulling plants the wool or lie skins is partly cleaned by pastrami them between revolving brushes upon which a stream of water runs. After pulling this partly cleaned wool is called brushed or brush.

**buck**
a male sheep (ram), goat, rabbit, deer, or antelope.

**bucket feeding**
feeding a young animal with milk and gruel from a bucket, rather than allowing it to suckle its mother.

- **bucket milker**

the type of milking machine where the cows are milked directly into a portable bucket.

- **buck wool**

the wool from rams. Buck wool can generally be distinguished by its characteristic odour. It is generally graded and sold separately. it is almost always heavier in shrinkage than corresponding eye wool-fine bucks frequently have coarse, smooth hairs mixed with the true wool fibre making the wool unsatisfactory to the manufacturer. For these reasons the manufacturers generally insist on buying buck wool at a discount, in the united states the wool dealers generally jay the growers only one-half lo two-thirds as much for buck wool as for wool from ewes, lambs or withers.

- **bulk classing**

a term used when quake wools of different brands and descriptions, but of similar type, yield, etc. are emptied out of their containers (bales), bulked together and rebaled tinder another or various brands into large links.

- **bulk collection**

assembly of similar materials from a collection into one or more bulks. Special bulks might be created of entries with some special attribute, i.e., resistance or equality trails. Actually, entries m any one bulk should be similar in origin, he'ght, maturity and adaptation.

- **bulk milk**

milk stored on the farm in a tank (or tanks), as compared with that stored in cans.

- **bull**

a sexually mature uncastrated bovine male.

- **bull dog calf**

calves born dead, usually prematurely; and the anatomical alterations, including short legs and bulldog like head arc common. There is excessive fluid in the foetal membranes. The placenta comes away id small fragments and there is no afterbirths in these bull dogs. A small metal appliance used temporarily for the-restraint of cattle. They are applied to the inside of the nose for holding an animal steady.

- **bulla**
  large sac like structure which is filled with fluid under the skin.
- **bulling**
  a cow in heat (estrus).
- **bulling heifer**
  a maiden heifer which has reached the right side and age for mating.
- **bullock**
  a castrated bull. A bullock calf is a castrated male calf.
- **bumble foot**
  a condition of poultry characterised by abscesses between the toes, caused by thorns, stones, glass fragments, etc, penetrating the soft tissue, or by bruising and rebutting in lameness.
- **bump**
  method of determining pregnancy during the just few months of gestation. The right side of the cow is humped sharply with the hand and the foetus swaying in fluid can be felt.
- **bunch**
  1. a handful of hay.
  2. a herd.
- **bunostomum**
  hoockwonns found in cattle, sheep and goat.
- **burdizzo emasculation**
  one of the commonly used emasculators. It has a compound lever which can he adjusted 10 that it locks firmly when closed.
- **burn**
  an injury caused by the contact of the skin with heal or by violent friction. The wound may cause redness of the skin or blisters. Severe burns may destroy soft tissues and even bones. Corrosive substance can also cause burn.
- **burry wool**
  wool that contains bum. It is undesirable because trouble and waste are always involved in the removal -of the burrs and they also add to the shrinkage. In many cases it requires a chemical process, called carbonising, to remove burrs completely.

- **bushel**
  a certain volume equal to 2150.42 cubic inches (approximately 1.25 cubic feet).

- **butcher system**
  a design, once popular, now rare, for hop garden wire-work, with three horizontal wires attaching to the poles, the lowest at about 15 cm, (6 in.) and the middle (breast wire) at about 120 cm. (4 ft.),

- **butenolide**
  a fungal toxin which can cause gangrene of the feet in cattle.

- **butt**
  a parcel of greasy wool in a recognised wool pack weighing less than 200 pounds gross (New South Wales). The weight varies in different slates; for example, in Victoria it is less than 225 pounds.

- **butterfat**
  the fatty substances contained in milk, mainly glycerides of palmitic and oleic acids, present as minute fat globules, the size of which vary according to breed and diminish from time of calving. The largest are present in jersey milk, churn easily and are best for butter making, the smallest are present in Ayrshire and Friesian milk and arc best for cheese. The butterfat content of milk is determined by the Gerber Test and now-a-days also by a number of semi-automatic instruments e.g., Foss Milkotester which are themselves calibrated using the Gerber Test.

- **butterfly**
  a breed of rabbit-butterfly value. A shut-off device where a disc is turned across the diameter of a tube.

- **buttermilk (dried)**
  the dried form of sour liquid left after the butterfat in milk ha1, been made into butter-butter milks. There arc two kinds of buttermilk, natural buttermilk is the by-product of churning cream into butter. Cultured buttermilk is usually made from fresh skim milk (cultured buttermilk may be made from reconstituted dry skim milk or reconstituted plain condensed skim milk). It is prepared by heating fresh skim milk to 85°c for 30 min-

utes (or equivalent), cooling to 21°c adding a lactic starter culture and incubating al 2 pc, until the acidity develops to about certain per cent. salt is usually added to accentuate the savour.

- **butyric acid**
  1. a fermentation product in silage making.
  2. the acid formed in bad butter; a colourless fatty acid with a rancid smell.
  3. a rancid, sticky acid from the putrefaction of protein; one of the volatile fatty acids produced by ruminal bacteria.

- **byre**
  a cow-nose.

- **cadaver**
  the body after death.

- **cade**
  a lamb or colt reared on the bottle.

- **caecum**
  the blind end of the large intestine, into which opens the termination of the small intestine.

- **calcification**
  is a term used id pathology to indicate that condition of a tissue in which there is a deposit of calcium carbonate laid down as a sequel lo an inflammatory reaction.

- **calcify**
  to deposit or secrete calcium sails that harden. An injured cartilage sometimes calcifies.

- **calcinosis**
  a pathologic condition in \which calcium in abnormal amounts has been abnormally deposited the condition of calcium deposits occurring in the skin is called *calcinco is cutis*

- **calcite**
  calcium carbonate, Ca.CO3 with hexagonal crystallisation, an acceptable source of car

- **calcium levulinate**
  while powder which is soluble in water and therefore used increasingly for the preparation of sterile solution.

- **Calcium Oxide (CaO)**
  quicklime. A lumpy while powder. It is caustic and irritant and is used to destroy carcases of animals, which have died from infectious diseases.

- **calcium pantothenate**
  commercially available salt of the vitamin b complex factor pantothenic acid. It is a white odourless, crystalline powder having sweet taste. It is soluble in water.

- **calcium phosphorus metabolism**
  the interchange of two important bone forming elements, calcium and phosphorus wit bin living organisms.

- **calcium phosphorus ratio**
  normal ratio of calcium to phosphorus in feed ratio has been between 2:1 add 1:1. the ratio 2:1 is generally preferred. Vegetable protein concentrates are poor in calcium and phosphorus.

- **calculus**
  a concretion formed in the body including gall-stones

- **Calf**
  the offspring (in its first year) of a cow. A male is known as a bull calf, a female is known as a cow calf.

- **calf crop**
  calves produced by a herd of cattle in one season.

- **calf diphtheria**
  a disease of calves generally under 6 weeks of age, in which greyish patches appear on the mucous membrane of the mouth and throat and less often on the nose caused by the Fusiformis necrophorus.

- **calf knee**
  the term is used for bone abnormality observed especially in horses.

- **calf pneumonia**
  it affects calves ranging from 15 to 75 days of age. The calf becomes weak due lo parasites and adverse environment. It causes a serious problem in many large dairy herds. It arises suddenly. The affected animal manifests dry cough, extreme lacrimation and a temperature range of 104.8x to 105.5°f.

- **calf scours**
  an important bacterial disease (often E. coli) of both dairy and beef calves, often fatal.

- **calf scours serum**
  it prevents the heavy losses from while scours if given to calves soon after their birth.

- **calk, calkin**
  a pointed, turned-down piece on a horse-shoe which prevents slipping.

- **callus**
  1. refers to the protective tissue which grows, as a swelling over wounds in woody plants.
  2. the cartilaginous growth, having collagen, which first grows to heal a bone fracture, later becoming hone tissue,
  3. any thickened tissue which is often hardened or leathery.

- **caloric**
  pertaining to heat or energy.

- **caloric energy of feed**
  the inherent power of feeds for supplying nutrients which are essential to the reproduction, maintenance, growth and milk production of animals.

- **calorie**
  the heat necessary to raise the temperature of 1 gram of water from 14.5 to 15.5°c is 1 (small) calorie; 1 (large) calorie raises the temperature of 1 kilogram of water 1°c. Thus, 1 Cal. = 100 cal; and 1,000 cal= 1 therm.

- **calorific value**
  of a fuel is the quantity of heal produced when a known weight of the fuel is completely hurl. It is measured in calories per gram or but per pound.

- **calorigenic**
  producing heat or energy

- **calorimeter**
  an instrument for measuring heat.

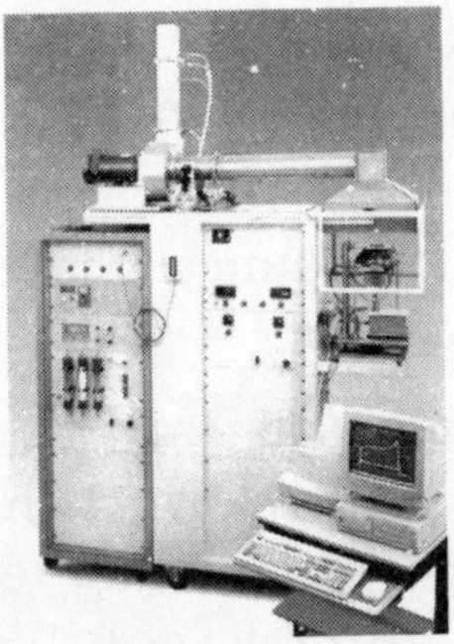

- **calorimetry**
  the science of measuring heat.

- **calve**
  to give birth to a calf.

- **calving interval**
  the average number of days between the latest two calving dates for all cows in the herd.

- **cambar**
an auto-sex- linked breed of poultry, with barred feathers, a single comb, yellow legs and producing white eggs.

- **can milk**
that milk plated in 10-gallon cans on the farm for cooling and transportation in a manufacturing plant.

- **Canadian bacon**
the bacon that results from mild curing and light smoking of pork sirloin muscle.

- **canaliculi**
minute canals or ducts which connect the lacunae in a bone.

- **canary stain**
a bright-yellow stain in the fleece which is not removed by normal scouring processes.

- **cancellated**
lattice- like- lace like- perforated with numerous holes or channels.

- **cancellous**
is a term applied to loose bony tissues as found in the ends of the long bones.

- **cancer**
a cellular tumour the natural course of which is usually fatal and often associated with formation of secondary tumours.

- **cancer eye**
it is a malignant tumour which, primarily attacks the eye of cattle. It may be caused by irrigation of the eyes by dust, sand, insects or strong rays of the sun. The growth has its origin in the *membrane nictitans* at the inner angle of the eye, in the mucous lining of the lid or in the cornea of the eyeball. Such growths are very sensitive and when injured bleed freely. Later, the sight gradually gel lost. Treatment involves the total removal of the cancerous tissue together with the eyeball by an experienced veterinarian.

- **candida albicans**
is a fungus which gives rise to the disease moniliasis or candidiasis: both in humans and in farm livestock.

- **candling**
the illumination of the egg interior by holding the egg before a light. This is possible because of the translucency

of the eggshell and the differences in the ability of other egg constituent to transmit light. Candling assures the elimination of practically all-inedible eggs.

■ **canine**

the family of animals which includes dogs, foxes and jackals, etc.

■ **canker**

1. a general term which is used for a number of fungal diseases of tree, especially those affecting fruit trees, in which growing tissues are destroyed and cortial tissues malformed, forming cancerous growths and open wounds.
2. fungus disease which causes softening of (he horn of a horse's hoof, inflammation and the production of a soft cheesy material,
3. a colloquial term for ear manage in dogs and for the pigmy conditions which can develop.

■ **canker, perennial**

a canker in which usually the intermittent activity of the pathogen or other injurious factor over a period of years prevents complete healing.

■ **cannery waste**

residue suitable for feeding from a canning.

■ **cannibalism**

the term used for the practice of an animal eating its own kind. it sometimes takes place, with indoor-housed poultry due to over- crowding, boredom (scratching for insects is precluded in cases) or protein deficiency and has been controlled by removing the point of the upper beak (do beating) or the use of soft red lighting. Sows will also occasionally eat their own or other sows' piglets under intensive conditions.

■ **cannon bone**

the large metacarpel bone in the foreleg or hinding of equines .

■ **cannula**

a device used to connect the rumen with the outside environment. The opening leading from the rumen to the outside is called a fistula. a metal-, rubber-, or glass-tube cannula may be inserted into a body cavity to allow the escape of fluids or gas liquids and other materials may be introduced into the body through s cannula,

- **canter**
the three-beat gait of a horse in which the feet strike the ground one at a time; an easy gallop.

- **cap**
dumb; also back of fowl's skull.

- **cap spinning**
a method of spinning in which each spindle has a metal cap to guide the yarn onto the bobbin.

- **capacity**
the ability of an animal to consume large amounts of feed. A large barrel and stomach often indicate an animal has good, capacity.

- **cape**
the feathers beneath the neck hackle, between the shoulders.

- **capon**
a castrated cockerel. Nowadays caponisalion is carried out chemically by injecting female hormones (oestrogens), which make the testes to regress. jt also produces a more tender carcase and stops cockerels crowing and fighting.

- **caponettes**
male chickens, that have had their reproductive organs made useless by the injection of an estrogenic hormone (stilbestrol). The testes of these animals decrease in sue and the secretion of testosterone is inhibited, which in turn results in a regression of the secondary sex characteristics (comb, wattles, ear lobes, mating instinct and crowing).

- **capped**
a term applied to bales from which the cap has been removed, according to regulations.

- **capped hock**
it is a firm swelling which occurs on the point of the hock of livestock. This blemish may be as large ox an apple or so small that it escapes online. It occurs as a result of constant irritation,

such as may be produced by rubbing or kicking the walls of he stable. The coalition is not considered serious, as it is not accompanied by lameness.

- **caplet**
  a disease of sheep (mainly old lambs and young sheep), generally those grazing hill areas in autumn and winter, considered to be due to phosphorus deficiency. It is characterised by lack of thrift and thinning of the skull bones. In bad cases sheep are unable to eat or close the mouth.

- **caprice**
  Pertaining to or derived from a goat.

- **capsule**
  1. a soluble gelatinous case having a dose of medicine, administered orally to an animal.
  2. the fibrous membrane which envelopes the various organs of hick body, e.g., liver, spleen.
  3. a type of hard, dry, dehiscent fruit, which is opening in various ways in, releases the seeds, e.g., poppy.

- **caput**
  head.

- **carbonates**
  these compounds are used as agricultural insecticides and sometimes cause accidental poisoning in animals.

- **carob types**
  very bury wools, usually short, from which the vegetable fault cannot economically be removed by mechanical methods.

- **carbohydrate**
  an enzyme that hydrolyses carbohydrates,

- **carbohydrates**
  materials consisting chemically of carbon, hydrogen and oxygen. The most important carbohydrates are the starches, sugars, celluloses and gums. They are so named because the hydrogen and oxygen are usually in the proportion to form water. They are formed in plants by photosynthesis; carbohydrates constitute a large part of animal food.

- **carbon dioxide**
  $Co_2$. an odourless, colourless gas resulting from oxi-

dation of carbon. It is formed in the tissues and excreted by the lungs.

- **carbonising**
  the process by which vegetable matter is extracted from wool by treatment with acid or some o her chemical agent which destroys the vegetable matter but does not destroy the wool fibre. Wool that requires this treatment for the burrs or seeds is called carbonising wool.

- **carbuncle**
  1. a lesion consisting of multiple cutaneous abscesses that are connected by sinuses.
  2. an acute suppurative inflammation of the subcutaneous tissues and the skin.

- **carcass**
  the major portion of a meat animal remaining after slaughter. Varies among animals, but usually the head and internal organs have been removed. Skin and shanks are removed from cattle and sheep.

- **carcass disposal**
  burning and burial arc the two ways of disposing dead animals. Burning is adopted wherever wood is plentiful. It completely destroys the infection. Carcasses are buried when wood is search. Burial often permits return of infection.

- **carcass residue (mammals)**
  residue which is obtained from animal tissues exclusive of hair, hoofs, horns and contents of the digestive tract.

- **carcinogen**
  any cancer-producing substance.

- **carcinogenic**
  cancer-producing tendency.

- **carcinoma**
  a malignant new growth made up of epithelial cells tending to infiltrate the surrounding tissues and to give rise to metastases.

- **carded woollens**
  a terra which certain writers in the united states apply to fabrics made from wool that has been carded but not combed, in order to distinguish them from worsted fabrics which are made from wool that is both carded and combed, or pre-

pared and combined-carded- woollens is synonymous with the trade term, woollens.

- **cardia**
is the upper opening of the stomach at which tne oesophagus terminates it lies close behind the heart.

- **cardiac arrest**
the cessation of cardiac couput and effective circulation of blood.

- **cardiac arrhythmia**
an alteration or abnormality of normal cardiac rhythm.

- **carding**
1. a teasing process which breaks down clumps of fibres into an even film by passing the material between rollers covered with closely set fine pins orpoints.
2. a term applied by the trade to wool suitable for use as raw material for the woollen system of yarn production.
3. a term sometimes used with reference to crossbred wools of less than about sis or seven inches in staple, which arc more economically prepared for the comb on the worsted card than on 'preparing' machinery.

- **cardiovascular**
pertaining to the heart and blood vessels, cerebral. Relating to the brain or cerebrum.

- **carina**
1. any keel late structure.
2. the prominent ridge on the breastbone of a bird.
3. a longitudinal projection on the glares of certain grasses.

- **carminative**
a drug, which relieves flatulence and colic by stimulating the movements of stomach and intestine.

- **carnivore**
any flesh-eating mammal] which has teeth and claws, e.g., dogs, cats, bears etc.

- **carnivorous**
eating or living on other animals.

- **carotene**
a yellow compound of carbon and hydrogen that occurs in plants; a precursor of vitamin a. alpha, beta and gammier carotenes may he converted into vitamin a in

the animal body.

- **carotene pigments**
orange or yellow pigments (yellow compounds) which through food, are deposited in animal tissues

- **arytenoids**
a group name for lip chromes or carotene like plant pigments (yellow compounds), which through food, get deposited in animal tissues.

- **carpal**
have or pertaining to the wrist.

- **carpet wool**
a coarse wool of pour quality, that usually comes from unimproved sheep. It is used largely in the manufacture of floor covering, but it is also used in the manufacture of coarse fabrics for wearing apparel.

- **corpus**
usually called the knee, the human wrist or corresponding part of an animal's forelimb containing the animal's carpel boons.

- **carrageenin**
a hydrophilic colloid extracted from the seaweed known as Irish moss. It is used as a stabiliser in frozen desserts and in evaporated milk. The functional group sulphate reacts with proteins to form a gel.

- **carriage**
the bearing pose or style of a fowl when standing or walking.

- **carriculture**
the study of goat- keeping.

- **carrier**
1. an animal possessing or earning the germs of a disease having itself recovered and showing no symptoms, but nevertheless able to pass the infection on to other animals.
2. an edible material which is used to facilitate the addition of micronutrients to a ration.

- **caries**
areas or tooth decay,

- **carrion**
dead or decaying flesh of man or animal.

- **carrion crows**
often cause injury to ewes and lambs, sometimes death and in addition they may transmit vibrionic abortion.

Crows may also kill piglets or injure older pigs.

- **carry**
  to bear, as a pregnant cow carries a calf-carrying capacity.

- **carrying capacity (wildlife)**
  the maximum number of a given wildlife species which in any given territory will support through the most critical period of the year,

- **carry-on-period**
  the stage in poultry rearing following branding and before chicks arc able to survive without assistance.

- **cardiovascular**
  pertaining to the heart and blood vessels, cerebral, Relating to the brain or cerebrum.

- **carina**
  1. a longitudinal projection on the glumes of certain grasses.
  2. the prominent ridge on the breast-bone of a bird.
  3. any keel like structure.

- **carminative**
  a drug which relieves flatulence and colic by stimulating the movements of stomach and intestine.

- **carnivore**
  any flesh-eating mammal which has teeth and claws, e.g., dogs, cats, bears etc.

- **carnivorous**
  eating or living on other animals.

- **carotene**
  a yellow compound of carbon acid hydrogen that occurs in plants; a precursor of vitamin a. alpha, beta and gamma carotene may be converted into vitamin a in the animal body.

- **carotenoid pigments**
  orange or yellow pigments (yellow compounds) which through food, are deposited in animal tissues

- **carotenoids**
  a group name for lip chromes or carotene-like plant pigments (yellow compounds), which through food, get deposited in animal tissues.

- **carpal**
  of or pertaining to the wrist.

- **carpet wool**
  a coarse wool of pour quality, that usually comes from unimproved sheep. it is used largely in the manufacture

of floor covering, but it is also used in the manufacture of coarse fabrics for wearing apparel.

■ **carpus**

the human wrist, or corresponding part of an animal's fore-limb containing the animal's carpel bones, usually called the knee.

■ **carrageenin**

a hydrophilic colloid extracted from the seaweed known as Irish moss. It is use as a stabiliser in frozen desserts and in evaporated milk. The functional group sulphate readies with proteins to form a gel.

■ **carriage**

the bearing pose or style of a fowl when standing or walking.

■ **cartilage**

a firm but pliant type of tissue forming portions of the skeleton of vertebrates. The proportion of cartilage in the skeleton of young animals is greater than in mature animals.

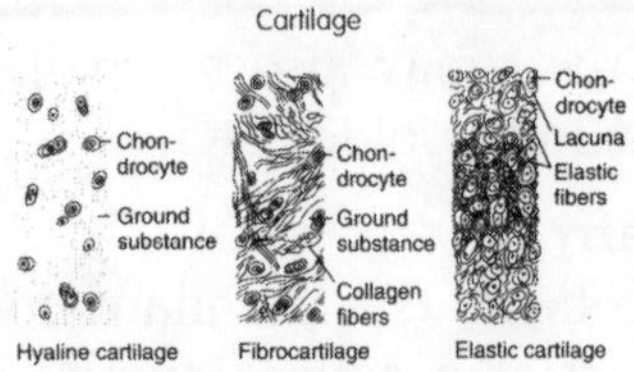


■ **catheter**

hollow tube designed to be inserted into various body cavities for the withdrawal of fluids or, in the case of artificial insemination catheter deposition of semen.

■ **catheterisation**

the passage of a catheter (a hollow cylinder through the urethra into the bladder) to drain urine in case of its retention. It is sometimes necessary in the treatment of sleeping sickness.

■ **cation**

an ion carrying a positive charge of electricity. Cations include the metal 5 and hydrogen,

■ **cationic**

action of a cation, for example, a detergent, such as ii quaternary ammonium salt, in which the/ cleaning action is inherent in the cation.

- **cattle**

bovine animals there are 250 major breeds and nearly breeds worldwide. They fall into two groups. Those mainly European breeds developed from *Baus Taurus* and those developed from *Bos Indicus*. Cattle breeds are classified as beef, dairy or dual-purpose.

- **cattle filarid**

a widely distributed nematode, found in the abdominal cavity of cattle.

- **cattle grid**

a type of grate comprising parallel bars in a frame covering it pit in the road. Usually used in replace gates. The spaces between the bars are wide enough to discourage stock from crossing but are no; a hindrance to humans or vehicular

- **cattle lice**

two types of cattle lice have been recognised. The suctorial lice are able to pierce the skin of the animal and suck its blood. The biting spies have got chewing mouth-part, which make them to feed directly upon skin tissues. Both species attach their eggs firmly to the hairs of cattle. Cattle lice could be conveniently controlled by dipping, e.g. nicotine dip, arsenic dip a sprayer or duster may also be used.

- **cattle lifter**

a stealer of cattle.

- **cattle plague**

a viral disease of cattle characterised by inflammation and ulcerations of mucous membranes. Eradicated from Britain m 1877. Still seriously affects cattle in Asia and Africa. Also called rinderpest or bovine typhus

- **cattle scab**

contagious skin disease which is affecting cattle of all classes and ages and is caused by mites which are parasites obtaining their food from the tissues of the host animal it can be eradicated by dipping or spraying lime-sulphur clip, petroleum dip etc.

- **cattle tick fever**

an infectious disease of the blood of cattle which is caused by the development and activity of minute protozon parasites which are conveyed to the animals by the cattle- fever tick. The disease has been characterised by high fever, destruction of red corpuscles, enlarged spleen, and enlarged liver, thick and flaky bile, bloody urine. Killing them on the pasture and on the cattle does the eradication of the licks. Animals could be immunised by introducing the casual parasite into the system by blood inoculation.

- **cattleyard report**
  day-to-day events happening in cattle yard are recorded which may include birth and death or treatment of sick animals as well as special wants like inoculation and other treatments.

- **caucdalalae**
  lateral extension of cuticle on mole tail.

- **caustic paste**
  prepared from one par arsenic trioxide mixed with five pans (lour and wet with water lo form a paste. this paste rinds use in the treatment of summer sores of horses.

- **cautery**
  the destructive effect produced by a cauterising agent used for scarring or burning the skin or other tissues by means of caustic chemicals or heat.

- **cavie**
  a hen-coop or cage.

- **cavy (plural, cavies)**
  one of the several species of tailless or short-lamed rodents of the family caviidae, like the guinea pig, capybara and mara.

- **cayenne tick**
  a parasite found on horse. Clipping helps to keep them off horses.

- **cayuse**
  Indian pony named after the Cayuse Indians, who currently live in Oregon,

- **cecum**
  an intestinal pouch located al the junction of the large and small intestine, also caecum.

- **cell**
  the cells are the structural units of which the body of the plant or animal is composed. It is a unit of independent mass of protoplasm with a nucleus in it, enveloped by a distinct cell

wall in the case of a plant but only a thin membrane (plasma membrane) in the case of an animal.

- **cell count**
the count of the number of white and red blood cells for blood corpuscles in a given quality of blood. It helps in the diagnosis of certain disease like anaemia.

- **cell division**
it is the process of formation of two daughter cells from a parent cell by enclosure of the two nuclei in separate cell compartments.

- **cell frequency**
when a frequency distribution is classified into categories, univariate or bivariate, the sub-categories are sometimes known as cells, the frequency with which observations fall into a particular cell is the cell frequency.

- **cell platelet**
a small, colourless, disk-shaped cell in the blood concerned with blood coagulation.

- **cell theory**
a theory enunciated by Schlyden and Schwan (1836-1839) which viewed the cell as the fundamental unit of life both structurally and metabolically.

- **cellulose**
the principal carbohydrate constituent of plant cell membranes. It is made available to ruminants through the action of microorganisms that inhabit the rumen (rumen bacterial flora).

- **celsius**
same as centigrade.

- **centaur**
a legendary monster of the Greeks with a weird combination of a horse's lower body and a man's head, arms and chest.

- **centigrade**
a thermometer scale in which water freezes at 0° and boils at 1xm)'. Same as Celsius. A means of expressing temperature.

- **central nervous system**
it comprises of brain and spinal cord.

- **centrifugal separation**
removal of one substance from another mechanically by centrifugal force. Materials to be removed must ex-

ist in a separate phase and one phase must have a higher density than the other, as with cream which is jess dense than skim milk.

- **centriole**
  a cell organ usually present as two small chromatic granules in the cytoplasm closely opposed to the nuclear wall. The mitotic centre in many cells. it plays a role in cellular division,

- **centromere**
  the point on the chromosome at which identical chromatids are joined and by which the chromosome is attached 10 a spindle fibre

- **cercaria**
  the final free swimming larval stage of a trematode, consisting of a body and tail.

- **cereal**
  cultivated members of the grass family whose seeds or rain arc used to provide flour fur bread making it as animal feed. The grain is rich with starch but also contains valuable proteins and vitamins. The Main cereals grown in India are barley, maize, oats, rye and wheat.

- **cereal forage**
  a cereal crop harvested when immature for either way- silage, green feed or as pasturage.

- **cereal substitutes**
  substances used by animal feed com-punier to provide the carbohydrate content of the feed instead of traditional cereals. The common substitutes include maniac, maize gluten, rice bran, milling offals, grain screen-mgs, citrus and beet pulp and brewing and distillery waste.

- **certified milk**
  milk produced and distributed under conditions which conform with high standards for cleanliness and quality set forth by the association of medical milk commission.

- **cerumen**
  the wax-like secretion of the specialised of the ear canal.

- **cervical**
  pertaining to the neck, the first region of the spinal column.

- **cervix**
  the neck as well as any neck like structure particularly the lower portion of the

womb (uterus).

- **cesarean section**
  delivery of a foetus through a surgical incision of the abdominal and uterine walls.

- **chafers**
  various beetles, the grubs of which are pests in old grasslands, usually those which arc in slightly dry situations.

- **chaff**
  glumes, husks or other seed covering combined oilier plant parts separated from seed in threshing of processing.

- **chaff cutter**
  a machine used for chopping fodder.

- **chaffing**
  the chopping up of straw and hay for feeding to stock. Chaff is commonly raised with concentrate rations for horses. it discourages hem from boiling the ration and compels them to chew it.

- **chalamyaspore**
  a thick-walled, intercalary or terminal vegetative cell which by thickening of the wall, has become modified into a resting spore.

- **chalazas**
  the spiral band of albumen joining either end of the yolk to the shell of a bird's egg.

- **challenge feeding**
  the concept is aimed a finding the potential of cows to secrete milk. Hence, an abundance of high-energy feed, balanced for the needs of high milk, production, are provided cows above what might apparently to be needed.

- **chapped teats**
  caused by any irritation like sudden chilling after sucking by the calf, wet milking, wet bedding for freezing in water. Treatment includes washing the udder with warm soapy water, rising and then drying with a towel and painting the chapped surface with glycerine iodine solution.

- **character**
  1. a specific recognisable feature transmitted from generation to generation, e.g., height, colour, etc in plant and animal breeding attempts are made to breed-in or breed-out characters

considered desirable or otherwise.
2. the nature and way of behaviour of an animal.
3. this term is difficult to define- it really applies to the possession of strongly marked distinctive 'qualities' of breed and type', the inbred or characteristic traits of wool. Well-bred wool of good character will generally show even, pronounced crimp and staple formation.
4. any well-defined feature, which distinguishes one species from another.
5. genetically, any feature which rest transmitted from parent to offspring.

- **check offs**

the practice of deducting a portion of the payment in a farmer for his produce, the deduced portion being used by a specific agency to promote safes of the specific product for example, check-off for milk may be transferred to the united dairy industries association for promotional uses of milk and milk products.

- **cheese**

a product made entirely from milk. In brief cheese making involves clotting milk by rennet, cutting the resultant junket or curd into small pieces and driving some of the whey out from the pieces of curd by the combined effect of hewing and acid development, the finished material betas packed into a metal or wooden container. The aim is to convert a considerable pan of the constituents of milk into a product that will keep well. Is less bulky, as it contains much lest water, is of high nutritive value and ix both pliable and readily digestible.

- **chelating agent**

an organic compound that can bind cations (metallic ions) by forming a stable, inert complex that t soluble m water; used in softening hard water, in purify nig sewage and in eliminating the concentrations of undesirable-metallic of radioactive elements- in the blood or tissues,

- **chemical analysis**

the determination of the composition of a substance using chemicals to break it down into simpler substances. The analysis may be qualitative, showing only the constituents present or quantitative, showing the actual amounts present. Used, for example, to determine the fertiliser requirement of soils, or the compositional quality of milk.

- **chemical score**
it is a method of determining the protein quality of teed in which the content of each essential amino acid is determined and expressed as the percentage of standard, the lowest percentage is taken in the score. For example in the case of wheat protein lysine is the first limited amino acid, the content of it is 7.2 and 2.7 per cent in egg arid wheat protein respectively. The chemical score for wheat protein is 2.7/7.2 x 100=37. These values compare favourably well with the biological values for protein in rats, pigs and human beings but not in poultry). It is because there are other acids, which are deficient in the protein and are not taken into account.

- **chemosis**
means swelling of the conjunctival membrane that covers the white of the eye, leaving the cornea depressed.

- **chemotherapeutic**
pertaining to the treatment of diseases by the use of specific chemical compounds diate tonic effects on the micro-organisms that cause the disease without seriously harming the patient. Penicillin, streptomycin, quinine and sulpha drugs are examples.

- **chevon**
meat of a goat kid.

- **chewing cattle louse**
the only biting species of cattle lice which concentrates, on the withers and around the root of the tail.

- **chewing the cud**
an activity of ruminants, involving the mastication for the second time of food which has previously been swallowed and passed in to the first stomach or rumen.

- **chiasma**
faint where chromalids in a bivalent join and where sub-

stitution of segments occurs during crossing over.

- **chick**
  a newly hatched fowl, or one that is not more than weeks old.

- **chicken**
  this is a word indefinitely applied up to the age of twelve months old.

- **chifox**
  a breed of ions-haired rabbit, variously coloured.

- **chigger**
  a larva (mite) infesting man, domestic animals, birds, snakes, turtles and rodents-its bite results in inflamed spots, accompanied by intense itching.

- **chill**
  1. to reduce the temperature of a product and increase its life and value.
  2. sensation of coldness, attended by shivering; may be followed by a variety of other disease symptoms in animals.

- **chipped rice, broken rice**
  the product comprising mostly small particles of rice kernels that have been separated from rest of the (comparatively) larger kernels of milled rice.

- **chipping**
  the process of cutting or breaking into fragments.

- **chitterlings**
  the smaller intestines of a pig or other animals sometimes used for food.

- **chlorine solution**
  a non-imitating disinfectant for washing udders of cows as well as teat cups of milking machines by dipping for one minute in a solution containing 200-300 ppm.

- **chlorine test (livestock)**
  the test used for determining mastitis. The test is based on the fact that normal milk is having a chlorine content varying from 0.08% to 0.14% and that the chlorides of the blood escape into the milk and the chlorine content of the milk has been raised with the inflammation of udder. It is a rapid test which should be run by a dairyman, but it should be used with a second test for confirmation of he findings.

- **chlorobutanol**
  it forms colourless to while crystals having camphora-

ceous odour which is commonly used as antiseptic, hypnotic and sedative. Heavy feeding of chlorobutanol may bring about paralysis and some time even death of the animal.

- **chlorocruorin**
  green respiratory pigment that occurs in blood plasma of certain worms.

- **chlorophyll**
  the green colouring matter present in growing plants.

- **choke**
  the stoppage of the gullet which is caused by an attempt to swallow large objects without thoroughly chewing them. The first symptoms are restlessness, salivation, retching, forced swallowing-rnovements and coughing. The object may sometimes become dislodged in 1 or 2 days without any mechanical assistance. Treatment consists in removing the obstruction causing the choke.

- **cholesterol**
  a white, fat-soluble substance found in animal fats and oils, bile, blood, brain tissue, nervous tissue, the liver, kidneys and adrenal glands. It is important for good metabolism and is a precursor of certain hormones.

- **choline**
  a colourless, viscid liquid which is a constituent of lecithin, choline is needed for normal growth and is sometimes called a *postulated vitamin* and is considered to have a function similar to that of true vitamin it is found in many seeds and with fatly acids, glycerol and phosphoric acid yields lecithin's, from which his liberated by hydrolysis. Chemically it is trimethylammonium hydroxide.

- **chondroblast**
  a cell which forms primitive cartilage.

- **chondroclast**
  a cell which breaks down cartilage.

- **chondrocyte**
  a cartilage cell.

- **chondrogenesis**
  formation of cartilage.

- **chopped**
  reduced in particle sat by cutting.

- **chopping**
the process of reducing the particle size by cutting with knives or other sharp instruments.

- **chops**
a mixture of dehydrated leaves and stems as they come from the dehydrator.

- **chorion**
the outermost membrane around a mammalian foetus (contiguous with the uterus surface) and around a developing chick in an egg.

- **chromatid**
a chromosome that appears doubled al the metaphase of cell division, conceded at one point by a small beadlike body called the centromere.

- **chromatography**
a technique for separating complex mixtures of chemical substances.

- **chromagenic**
capable of producing coloured pigments.

- **chromomere**
dense granules of chromatic material which appear on the chromosomes during the early stages of meiosis. They are characteristic in number, position and size.

- **chromosome**
dark-staining rod like or rounded bodies visible under the microscope in the nucleus of the cell in the metaphase of cell division. Chromosomes occur in pairs in body ceils and the number is constant for a species. Chromosomes carry genes arranged linearly along the chromosomes are the backbone of a chromosome is the DNA molecule.

- **chromosome map**
a diagram of chromosome showing the position of the genes. It is obtained by a statistical analysis of the frequency of crossing-over.

- **chromosome set**
the whole complement of chromosomes. Present in, or derived from the nucleus of a gametes.

- **chronic**
of long duration as opposed to acute.

- **chuni**
chuni means churna. The compound consists primarily of the broken pieces of en-

dosperm including germ and a portion of husks obtained as by-product during the processing of pulse grains for human consumption, e.g., arhar chuni, masur chuni, mung chuni etc are the compound is valued, as a concentrate feed due to comparatively low in fibre and more in energy and protein content in comparison to roughage.

- **churn**
  1. a large metal milk container or can, commonly of 10 gallons capacity (occasionally 12 gallons), with a close-fitting mushroom type lid.
  2. a machine which agitates cream or whole milk turning it into butter.

- **chyme**
  a thick liquid of partially digested food- it passes from the stomach into the small intestine.

- **CIP (cleaned in place)**
  most stainless steel pipelines are cleaned by circulating washing solutions through them.

- **circling disease**
  a specific bacterial disease affecting mammals including man and birds. It is characterised by meaning-encephalitis and sometimes abortion in domestic ruminants and human beings. It is caused by listeria monocytogenes, which is a small, rod shaped, 1- long and 0.5 to 0.6 broad, gram positive and non-sporing bacteria.

- **circulation cleaning**
  a means of cleaning the internal surfaces of a milking machine that are in contact with the milk by circulating the clean fluids.

- **cirrhosis**
  a chronic disease of the liver characterised by a destruction of liver cells and an increase of connective tissue

- **cistern**
  area in the udder into which the milk is secreted before entering teat.

- **cistern milk**
  milk, in the lower cistern cavity (if the udder which can easily be withdrawn without stimulation.

- **citrated blood**
  the blood drawn from a healthy animal into a 2.5%

to 4.0% w/v sodium citrate solution which prevents coagulation when used at the rate of i cc/10 cc of blood. this provides a transfusion mixture containing 0.25% to 0.4% sodium citrate.

- **citrulline**
  one of the don-essential amino acids.

- **claret**
  the breast colour of a Rouen drake.

- **clarification**
  the process of removing suspended material; in milk, processing, clarification is achieved by centrifugal realmerit which removes sediment (extraneous matter) and somatic cells.

- **class I base plan**
  the system used in some federal milk marketing orders whereby a producer is allowed to deliver a specific quantity of milk at the class i price hut il penalised for deliveries in excess of that quantity.

- **class I, class II, class III**
  the United States tariff laws from shortly after the civil war irntrm9l2dividtd foreign wool into three classes for the purpose of assessing duties. Class I included all wool from merino sheep and crossbred sheep containing merino blood; class it, the medium arid tong woofs from British sheep and class iii, native, unimproved wools from all over the world, commonly called carpet wools. Although recent tariff laws use a different classification, the old terms are still used lo some extent in the literature of the wool trade especially in market quotations

- **class I milk**
  milk used in fluid milk products. Dairymen for this class of milk receive the highest price.

- **class II milk**
  milk in excess I needs, consequently used in a lower price category. This includes milk used for manufacturing purposes. In some markets, surplus milk is divided into two classes. Class II and class III and milk is allocated to the respectively classes based on use.

- **class I price**
  the minimum price that handlers must pay for milk

used in fluid milk products in a federal order market.

- **class II price**
the minimum price handlers must pay for milk used in the class ii category; usually the price equals the price of manufacturing milk in Minnesota and Wisconsin.

- **classification score**
a final point rating placed on animals in the breed association conformation rating programmes according to how closely each animal resembles the breed's ideal conformation.

- **classified pricing**
a pricing plan by which handlers are charged for market milk according to the products in which it is used, hence the term used classification. Class I milk goes into fluid products and demands the highest price. Prices of surplus milk. Class II or iii, depending on the marketing order, approximate the price of manufacturing milk.

- **classified pricing plans**
a schedule of prices which handlers pay for milk depending on whether it is used as fluid milk or for manufacturing purposes.

- **classing**
1. of stud ewes-the division of a flock of stud ewes into various groups prior to mating.
2. of sheep— the process of culling and selection applied to a flock of sheep.
3. of wool— the preparation of the clip for market the wool is graded into even lines according to type, yield and commercial value.

- **clat**
to remove wool from around the udder and the inside of the thighs of a ewe prior to lambing.

- **clavicle**
the collar bone.

- **claw (of milking machine)**
a chamber to which the teat cups arc attached delivers milk from the teat cups to the milk line.

- **clawpiece**
sometimes abbreviated to claw. Contains a milk chamber, pulse chambers and connections lo the long milk tube, long pulse tube liners and shells.

- **clean legged**

  a descriptive term for a horse with few or no long hairs (known as feather) on its legs.

- **clean pasture**

  the sward after an arable crop, following receding, ungrazed by livestock and therefore presumed to be uncontaminated by parasitic worm larvae.

- **cleft**

  split or divided to a certain depth, a cleft palate is a longitudinal opening in the roof of the mouth. A cleft-footed animal, also called cloven footed. E.g., the cow or hog has a divided hoof or foot.

- **cleft palate**

  hereditary defect in pigs. it is characterised by their inability to nurse, they die soon after birth.

- **cleg**

  one of various species of blood sucking, two winged, tabanid flies (*haematopotta sp.*). They arc ashy-grey with finely mottled wings and large bright eyes and arc troublesome to horses. It is also known as horse fly.

- **cleidoic**

  dosed or lacked in, as the egg, which is cut off or isolated from free exchange with the environment by more or less impervious shells.

- **climate**

  the prevailing weather conditions within a notified area.

- **climatic cycle**

  actual or supposed cyclical recurrences of such weather phenomena as wet and dry years, hot and cold years, at more or less regular intervals, in response to long-range terrestrial and solar influences, such as volcanic dust and sunspots.

- **climatic factors**

  include temperature, precipitation, light, evaporation, air, radiation, humidity, atmospheric pressure and winds that affect the biosphere.

- **climatic region**

  an area on the earth's surface over which the combined effect of climatic factors result in an approximately homogenous, set of climatic conditions. it is referred as a climatic type.

■ **climatic year**
a continuous twelve-month period during which occurs a complete annual climatic cycle and which is selected for the presentation of data relative to meteorological phenomena,

■ **climatography**
the adaptation of a species to a particular climate.

■ **climatology**
the science that seeks to describe and explain the nature of climate, why it differs from place to place and how it is related to other elements of the natural environment and lo human activities. The term has been derived from Greek words 'klima' meaning slope or latitude of earth and 'logos' meaning study.

■ **climax**
the final or stable community in a successions series, which is self-perpetuating and in equilibrium with the physical habitat,

■ **cling**
diarrhoea in animals, particularly sheep.

■ **clinical**
the investigation of a disease on a living subject by observation, as contrasted to a controlled experiment.

■ **clinical abnormalities**
unusual condition or behaviour observed in a diseased animal. It is used as an aid in diagnosis.

■ **clinical evidence**
any symptom of disease that can be determined by direct observation such as lack of appetite,fever , swelling etc.

■ **clinical infection**
an infection in which clinical symptom develop.

■ **clip**
1. removal of hair from the cow with shears especially around udder, tail etc.
2. one of several projections which prevent a horse-shoe from shifting on a horse's foot.
3. the wool removed from either a single sheep or a whole flock.
4. to remove or shear the wool from a sheep.

■ **clipped**
with oat grain, the more fibrous end has been removed.

▪ **clitoris**
a small organ situated at the interior port of he valve.

▪ **cloaca**
the common cavity of the chicken through which the reproductive and the digestive tracts discharge their products from its body.

▪ **clone**
a group of plants or animals all having an identical genetical make up, being the descendants of a single parent. In plant taking cuttings from a single parent plant can vegetatively raise breeding a clonal slock of plants.

▪ **clonical**
it is a word applied to spasmodic movements of muscles lasting for a short time only.

▪ **clonorchiasis**
oriental liver fluke disease (especially in China and Japan), the causative parasitic trematodes have two intermediate hosts; the first is a molluscan and the second is some edible fish from which man becomes infected.

▪ **close**
descriptive of body and texture. In frozen desserts, it is the quality imparted by having minute ice crystals and air cells so the body is smooth (the opposite of coarse). In cheese, it is the absence of mechanical openings or gap holes (the opposite of open').

▪ **close breeding**
1. a form of inbreeding, e.g., the mating of brothers to sisters, sire to daughter and son to dam.
2. an intense form of inbreeding usually applied to the mating of closely related animals such as full sibs or parent-offspring.

▪ **close season**
the period during which it is declared unlawful under an act or ordinance to hunt. Catch or kill any particular animal, bird or fish.

▪ **closed herd (rock)**
a herd (flock) in which no outside blood is introduced.

▪ **clothing wool**
wool that is presumed to be too short to comb. In the United States trade the term refers to length alone as is shown by the fact that carpel woof is divided into

clothing and combing wool, according to whether it is long or short. In Australia the term clothing is applied only to short-stapled line

- **cloudburst**

it is a colloquial name for false pregnancy in the goat which, after an apparently normal gestation, suddenly voids from the vulva a large quantity of cloudy fluid after which the size of the abdomen returns to normal. Cloudburst is a fairly common condition.

- **cloven hoofed**

a descriptive term of animals with a divided-hoof such as cattle and sheep.

- **clover**

a large genus (*trifolium*) of leguminous plains characterised by trefoil leaves, having small flowers with light roundish heads and producing one to four-seeded pods. root nodules contain bacteria capable of nitrogen fixation. A number of grass seeds to produce leys, e.g.. Alsike clover (*T. hyble*), Wild while clover or Kentish clover (*T. repens*), purple or red clover (*T. pratense*). Clovers usefully knit he sward together and help keep out weeds, especially in longer

- **clover hay**

hay derived from a clover crop. Also hay from a temporary Ley as opposed to that derived from permanent grassland (meadow hay). Also known as seeds hay.

- **clover sickness**

a condition of a field on which clover has been grown for loo long and which is heavily infected with ele-worms and fungus diseases.

- **cluster**

1. the four teat cups of a milking machine applied to a cow's udder.
2. the leaves, flowers or fruits growing from a common point on the stem or shoot of a plant.

- **clutch**

the eggs laid by a hen or consecutive days are referred to as n clutch. in domestic birds the number of eggs laid successfully are often referred to as a cycle or laying, or the laying rhythm, the term clutch for wild birds refers to the set of eggs laid for one incubation. Some birds lay a single egg (penguin); others lay 2 (pigeon); others may lay 12 to 20 (partridges and most domestic fowls).

- **coagulant**
  a substance that acts on a liquid to coagulate it.

- **coagulated**
  curdled, dotted, or congealed,

- **coagulation**
  the change from a fluid slate to a thickened jelly. Curd, or clot.

- **coarse**
  a term used in evaluating dairy products. In frozen desserts, it denotes lack of smoothness; in cottage cheese and buttermilk, it means an overabundance of diacetyl, in butter, a lack of sweet, pleasant, delicate flavour.

- **cobalt**
  it is one of the mineral elements known to be essential to normal health, hut only required id minute amounts. Because of this, cobalt is said to be one of the 'trace elements'. Its function is to act as a catalyst in the assimilation of iron into haemoglobin m the red blood corpuscles.

- **cobalt deficiency disease**
  disease caused in animals due-to cobalt-deficiency; showing paleness of mucous membrane such as conjunctiva; symptoms of anaemia, reduction in milk yield in cows, diarrhoea in calves, treatment by feeding with sulphate or chloride of cobalt.

- **cobby**
  squarely built, short and blocky in body-size.

- **cobweb theorem**
  an economic theorem that producers of a commodity expect the price in he next period of production to be the same as current prices and thus plan production accordingly. It is most appropriate in the ease of non-storable commodities.

- **cocaine**
  an alkaloid obtained from the leaves of *erythroxylum coca*. It finds use as a local anaesthetic and mydriatic. Its solution in water is some-

times applied to soothe the pain in eyes.

- **coccidioidal granuloma**
the non-contagious fungus disease of animals, which is characterised primarily by multiple pulmonary granulomas and dissemination to other organs. Coccidioides inmitis is the casual organism,

- **coccidiosis**
disease, mainly of lambs, due to infection with a microscopic protozoal parasite, especially under overcrowded conditions. Causes extensive cell destruction and hence weakening of lambs, which may die from exhaustion.

- **coccidiostats**
drugs used to control coccidiosis and usually given to poultry in drinking water or, mixed in the feed.

- **coccygeal**
pertaining to the tail-posterior— the fifth region of the spinal column.

- **cock**
a general term for a male bird. For poultry, usually applied to males over is months old.

- **cock hen**
an apparent change of sex in which a female assumed male characteristic about the head. Due to inhibition of the female sex hormones.

- **cockerel**
in show parlance, a bird that is hatched in the year it is exhibited. in utility language a young male until the first adult moult.

- **cockerel bred**
birds from a pen specially mated to produce cockerels,

- **cocksfoot**
a perennial grass (Dactylis Glomerata}, with characteristic one-aided flower-heads of rounded or oval individual flower spikelets on wiry branches and dull green to deep blue leaves. Deep rooting. Very high yielding, used extensively for pasture.

- **cocoa**
the product derived from fermented bean of the cocoa tree. Fermented beans are crushed and the nibs are separated to be hydraulically pressed. Pressing removes fats and forms a cake, which is ground to

produce the cocoa.

- **cocoa poisoning**
poisoning of pigs and poultry, as a result of feeding cocoa residues or waste.

- **coconut cake**
a concentrate feeding stuff reasonably rich in protein, used frequently in rations for dairy cows, fattening - cattle and pigs. It produces a harder fat than most other oil seed products. Not as readily available as it used to be.

- **coconut oilcake**
the product is obtained after removal of oil from dried coconuts by a mechanical extraction process.

- **cod**
the part of the scrotum left after castration.

- **cod liver oil**
oil obtained from the livers of cod-fish. it is yellow in colour and produces a disagreeable odour on exposure to air, it is used as s tonic and nutrient. it ii one of the most valuable sources of vitamins a and d available for animal feeding. it is particularly helpful m preventing rickets developing.

- **coefficient**
an expression of a relationship between two kinds of observations. Thus, the speed of a car in miles per hour is a coefficient. It gives the relationship between miles travelled and hours elapsed.

- **coefficient of determination**
the square of the multiple correlation coefficient, $R^2$, is referred to as the coefficient of determination,

- **coefficient of digestibility**
the percentage (if each of the food constituents present in the feed that are digested by the particular kind of animal. the digestibility of a certain feed may be different for different kinds of animals and also for different ingredients of food.

- **coefficient of variation (cv)**
the percentage variation in the mean, the standard deviation being treated as the total variation in the mean.

- **co-enzyme**
a substance, usually a vitamin or mineral, which

works with an enzyme to perform a certain function.

- **coffee meal**
ground coffee beans after extraction of oil.

- **coffee pulp**
the outer covering of the coffee berry. It bat got about 17.5% protein but poor palatability.

- **coffin joint**
the joint located between the second and third phalangel bones of the hoof.

- **coggin test**
a teat used to detect anaemia in horses.

- **coherent**
united, but very slightly, so that joined organs appear free: when viewed superficially.

- **coitus**
sexual intercourse between individuals of the opposite sex.

- **cold blooded**
poikilothermic; without ability to regulate the body temperature by physiologic means: commonly said of fish, reptiles and amphibians whose temperature approximate that of the environment.

- **cold branding**
branding of farm animals using caustic depilatories, which produce a permanent skin brand without discomfort to the animals.

- **cold brooding**
also known as hay box brooding. It is a. simple method of rearing of chicks when it is not too cold, by employing an ordinary basket or small box lined thickly with fine hay or old flannel strips of feathers and provided with small hole or outlet to allow passage to chicks.

- **cold sterilisations**
the use of a cathode ray or an electron-beam gun in food processing to kill bacterial or insect life. Chemical sterilisation of instruments.

- **colic**
a digestive disturbance causing pains in the abdomen, as in horses.

- **coliform bacteria**
those bacteria commonly present in animal intestines

and passed out in faeces (e.g., *escherichia coli*). The presence of such bacteria in milk indicates faecal contamination.

- **coliform test**
  test of estimating coliform bacteria in a sample of milk, presence of such bacteria being an index of unhygienic method of milt production.

- **colitis**
  the inflammation of the large intestine which is often due to diarrhoea, constipation, etc.

- **collagen**
  a proteinous substance, the principal fibrous constituent of akin, tendon, ligament and bone. When boiled it produces gelatin.

- **collar**
  a white ring round the neck Rouen duck.

- **collateral relatives**
  those related individuals that are not related as ancestors, or descendants.

- **colligative properties**
  characteristics that vary with the number of molecules but not their nature. Freezing point, boiling point osmotic pressure and. vapour pressure are colligative properties of solutions.

- **Collings Brothers**
  British cattle breeders who successfully applied many of bake welts principles plus their own insight in a manner, which led to the formation of the shorthorn breed and establishment of the first handbook for cattle.

- **colloidal**
  in a state of fine sub-division or dispersion with particles ranging between $10^{-5}$ and $10^{-7}$ centimetre in diameter and having no effect on the colligative properties of solutions, the casein of milk is in colloidal suspension.

- **colon**
  the greater part of the large intestine exceeding from the cecum to the rectum.

- **colony brooders**
  brooders which are used for rearing chicks in jots from about 60 to a few hundred, with kerosene oil, coal, wood, oil, gas or electricity supplying the fuel.

- **colostral milk**
first milk which is secreted soon after parturition. colostral milk helps in cleaning out the digestive trait and provides the newborn animal resistance against harmful bacteria.

- **colostrum**
it is the milk secreted by the udder immediately after parturition and for the following 3 to 4 days. It consists. of the cells (epithelia) which previous to the birth of the young animal were picked in the acini, mixed with serum and fat globules. sic. It contains 20 pr cent or more protein, a little more fat than normal milk and may be tinged pink due to blood corpuscles.

- **colt**
a young horse; foal. male horse under four years of age. (a young female horse is called a filly).

- **coma**
it is a state of profound unconsciousness in which i be patient not only cannot be roused, but there are no reflex movements when the skin is pinched or pricked, or when are eyeballs are touched, etc.

- **combine ability**
the ability of two inbred lines or any two stocks to cross and produce superior offspring.

- **comb**
the fleshy ornament on the head of a fowl there arc kinds of comb, rose, pea, etc.

- **comb over**
having the comb dropping over to one side as in leghorns, Minorca, or heavy rose combed birds, or when the comb does not sit squarely on the head,

- **combing**
an intermediate process in the preparation of wonted yarn, which renders the longer fibres parallel and removes, as soil, short fibres, broken ends, tangled neps and larger vegetable particles.

- **combing wool**
a wool that has good length and is long enough to comb. the U.S. has established a table of standard minimum lengths for wool that is to be classified as strictly combing. according to this table, the

length of staple for fine strictly combing must be over 2 inches and increases 1/4 inch for each grade as the wool gets coarser.

- **combining ability**

the relative ability of an inbred or a clone, when crossed in another inbred line or clone, to transmit desirable traits or a specific trait to its cross. It is of two types,

1. specific combining ability h used to indicate those crosses in which certain combinations to relatively better or worse than would be expected on the basis of the average performance of the lines involved'.

2. general combining ability is the average performance of a line in hybrid combination and

- **combustion**

the combination of substances with oxygen accompanied by the liberation of heat.

- **comeback**

1. the kind of wool that is ordinarily produced by a comeback sheep. it is somewhat coarser than the grade called half blood in the united states. one Australian author says that it is of 58's and 56's quality. another states that some comeback fleeces contain 64x5 wool.

2. an Australian term meaning a sheep that has a crossbred dam and a merino sire.

- **comedo**

a plug of keratin and dried sebum in ail excretory duct (or hair, follicle) of the skin.

- **comfort zone**

the temperature interval during which no demands are made on the temperature regulating mechanisms. the temperature at which man and animals feel most comfortable.

- **commensal**

living on or within another organism and deriving benefit without injuring or benefiting the other individual. not parasitic symbiosis.

- **commercial feed**

any material produced by a commercial company and distributed for use as a feed or feed component.

- **common stomach worm**

a nematode found in the fourth stomach of cattle, gout and sheep. it is 1 to 3 cm. long and as thick as an ordinary pin. This blood sucking parasite is responsible for the loss of flesh, general weakness, anaemia and diarrhoea. infested animals may be treated with copper sulphate or phenothiazine.

- **common wool**

the next to the coarsest grade in the united state official wool grades established by the u.s. bureau of agricultural economies. In the United Slates in the early part of the nineteenth century it was the name given to the rather coarse wool produced by the common sheep of that day as contrasted with the fine wool produced on the newly imported merino sheep.

- **communicable disease**

a disease whose causative agent is readily transferred from one person toanother.

- **community (ecology)**

a collection of population of animals and plants that occurs naturally together in a common environment, such as the organisms inhabiting a woodland. The individuals and populations within the community interact with one another and with the abiotic (non-living) surrounding, the community and abiotic surroundings together consenting an ecosystem. An open community, is one into which new species can be absorbed; a closed community is one in which no other species from outside the community can interact with those within it.

- **compaction**

closely packed feed in the stomach and intestines of an animal causing constipation and/or digestive disturbances.

- **compatible**

two plants whose leaf edges have a permanent North-South orientation

- **compensation**

the method by which the body makes good a defect of form or function in an organ which is abnormal in these respects.

- **compensatory growth**
  the phenomenon of animals turned out to pasture after a period of energy restriction exhibiting greater growth rate than animals which have not been so restricted. It is due to the greater feed intake following the period of energy restriction.

- **complement**
  a protein which belongs to the normal thermolabile group and present in plasma and tissue fluid. It participates in antigen-antibody reactions.

- **complement fixation**
  this is a biological reaction that is the basis of many tests for infection. It utilises a three-phase system:
  1. antigen,
  2. complement and
  3. antibody.

  when a specific antigen and antibody are mixed in the presence of complement, inactive or fixed. Therefore, if the presence of a specific antigen or antibody in a mixture is known, the presence of the other can be determined by the mixture's reaction on the complement. Such a reaction is the basis of many tests for infection, including the Wassermann test for syphilis and reactions for gonococcus infections, glanders, typhoid fever, and tuberculosis.

- **complementary genes**
  genes that so interact that when both are present a new or novel trail appears.

- **Complementary milk**
  portion of milk present in the udder that is not removed even by thorough milking.

- **complete diet**
  winter feed for livestock comprising a mixture of concentrates and bulk forage fed as a complete mixture, the animals usually receiving no other feed.

- **complete feed**
  a nutritionally adequate feed for animals other than man, is compounded by specific formula to be as the sole ration ,and is capable of maintaining life and/or promoting production without any additional substance being consumed except water.

- **complete ration**
  a blend of all feedstuffs

(forage and grains) in one feed. A complete ration fits well into mechanised feeding and the use of computers to formulate least-cost rations.

- **component milk pricing**
payment for milk on the basis of composition partially in addition to milk fat. Usually solids not fat or protein content.

- **composite hypothesis**
a statistical hypothesis which is not simple and in practice, refers to hypothesis which is composed of a group of simple hypothesis.

- **compound**
a term used by corn merchants to denote a mixed mash.

- **compound feed**
animal feed composed of several different feeding stuffs (and including major minerals, trace elements, vitamins and other additives) in proportions providing a balanced diet. the compound manufacturers provide a range of feeds suitably balanced for all types of stock at all growth and development stages.

- **concentrated milk**
the liquid food obtained by partial removal of water from milk. The MF and TMS are not less than 7.5 and 22.5 per cent, respectively. It is pasteurised but not processed by heat so as to prevent spoilage. It may be homogenised. Also called *condensed milk*.

- **concentrates (feed)**
1. feeds hat are low in fibre and high in total digestible nutrients. technically, all feeds supplying primary nutrients (protein, carbohydrate and fat are classed as concentrates if their crude fibre content is less than is per cent. examples are the grains, cottonseed meal and wheat bran.
2. a fortified supplement feed which requires the addition of other feed ingredients to form a complete balanced ration for animal, fowl or fish. concentrate feeds are capable of sustaining life, but generally require the addition of other feed ingredients to obtain or sustain maximum production.

- **conception**
the fecundation of the

ovum. The action of conceiving or become pregnant, it is the first stage in the development of an embryo.

- **conception rate**

in cattle breeding, the percentage of first: services that conceive.

- **conceptus**

the product of conception; initially a fertilised egg, later an embryo which develops into a foetus plus foetal membranes.

- **condensed milk**

enriched milk produced by evaporating much of its water content and by the addition of sugar,

- **condensing**

dividing the web of carded wool into strips and rubbing them into the form of slubbings.

- **condition**

1. the amount of flesh (body weight), the quality of hair coats and the general health of animals.
2. the state of a bird's health, or plumage, especially the latter.
3. the general health, cleanliness of plumage and brightness of headgear.
4. the degree of fatness of meat animals.

- **condition wool**

1. the state of wool with respect to the amount of yolk and foreign matter of wool it contains. Wool that is heavy in condition will shrink heavily when it is scoured and wool that is light in condition vice versa.
2. the amount of moisture present in wool, expressed as a percentage of clean dry weight.
3. wool- a qualitative expression of the amount of yolk and other impurities present in the raw wool.

- **conduction**

a means of dissipating heat, especially on contact with cool water- the transfer of heat.

- **condyle**

a cylindrical articular surface.

- **condyloid**

resembling a condile.

- **confluent**

1. becoming merged or running together;
2. not discrete.

- **conformation**
  the physical form or physical traits of an animal; it is the shape and arrangement of parts.
- **confounded and fractional replication**
  a factorial experiment in which only some of the treatment combinations occur in each block. Particularly used when several factors each at 2 or 3 levels are present.
- **confounding**
  a situation whereby two or more causes of variation exert simultaneously their influence on the experimental material making it difficult or even impossible to ascertain the exact cause of the observed effect (in statistics).
- **congenital**
  deformities, diseases, etc., are those which are either present at birth or which, being transmitted direct from the parents, show themselves some time after birth.
- **congestion**
  excessive accumulation of blood in a part of the body.
- **conjugate**
  a substance formed by the union of two compound.
- **conjugated proteins**
  proteins which are united with some non-protein substance by a linkage other than a salt linkage.
- **conjunctive**
  membrane that lines the inner surface of the eyelids and covers the frontal portion of the eyeball.
- **connective**
  a thing that is used to join various parts of a structure.
- **connective tissue**
  a type of supporting tissue which is made of protein and is especially strong. Some appear as long as distinctive strands; others as a cementive mass between muscle cell.
- **consolidation**
  a term applied to solidification of an organ, especially of a lung.
- **constipation**
  the retention of faeces in the intestines because of difficult evacuation. It may be developed due to feeding of dry,

coarse or indigestible feed or due to overeating or from overfeeding.

- **constitution**
  the general bodily vigour indicated by large heart girth roomy middle and rugged bone.

- **constitutional**
  relating to the system of the body as a whole.

- **contact animal**
  an animal which has been in contact with a diseased animal. They are needed to be isolated if the disease is infectious.

- **contact dermatitis**
  inflammation of skin resulted from external application of a primary irritant

- **contact insecticide**
  any substance that can kill insects by contact

- **contagious disease**
  a kind of disease that is conveyed by actual contact from a sick animal to another.

- **contamination**
  entry of unwanted substance or micro-organisms to some material or object.

- **contemporary comparison**
  a method for estimating the ability of transmission of genetic material of bulls and cows using information on contemporaries, raised and completed their lactations during a certain time periods frequently limited to first lactation records.

- **contemporary group**
  a grouping of records into 2 groups, first lactations in one group and all other lactations in another group.

- **contingency table**
  where the characteristics of the members of an aggregate are quantitative, classification may be sent out in a two-way table known as 2x2 *contingency table.*

- **continuous brooders**
  brooders used for rearing chicks.

- **continuous distribution**
  method of depicting the frequency of occurrence of different levels of traits which do not occur in discrete classes (such as sex) but vary in increments too small to mea-

sure such as body weight. Such traits are usually measured for convenience in arbitrary finite units such as pounds, kilograms, inches, or centimetres.

- **continuous grazing**
grazing animals on a single pasture area continuously the whole the pasture season.

- **continuous variation**
1. variation shown by such slight individual difference between member specimens belonging to the same lineage that one specimen may shade into another.
2. a series of minute variations;

- **control, biological**
1. the artificial application of biotic control .
2. the use of the diseases and enemies of a pest for the purpose of maintaining adequate control.

- **control, biotic**
the natural control of the numbers of a plant, animal, or insect by its diseases and enemies.

- **control rod**
a rod, plate, or tube used to control the power of a nuclear reactor. It contains a material that readily absorbs neutrons. By absorbing neutrons, a control rod prevents the neutrons from causing further fission.

- **controller value**
the valve arrangement inside a vacuum controller.

- **convection**
the method of either warming or cooling the animal by breezes.

- **convergence**
morphologic similarity in distantly related forms.

- **convulsion**
a violent involuntary contraction or series of contractions of the voluntary muscles.

- **cool season grass**
a particular grass species adapted to rapid growth during the cool, moist periods of the year.

- **coop**
a small wooden structure for keeping hen and chickens or fowls to fatten, or for the suppression of broodiness.

- **copper**
one of the trace elements

which is essential in the nutrition of animals, acts as a catalyst in the assimilation of iron, which is needed in the production of haemoglobin m the liver. Its absence from the foodstuffs eaten in some areas leads to a form of anaemia.

- **copra meal**
  the remains after the dried edible portion of cocoanuts have been subjected to fat extraction and ground. The relatively low quality of its protein and its relative high content of fibre restrict its use in rations for swine and poultry.

- **coprophagy**
  the eating of its own faeces by an animal.

- **copulation**
  sexual union of two animals with the transference of sperms from the male to the female body

- **core**
  1. pierce the body with an animal's horns.
  2. the central portion of a nuclear reactor, containing the fuel elements.

- **corn stover**
  the dried cornstalk from which the ears are removed.

- **cornicle**
  an abortive spur on a hen's leg which becomes harden with the increase of age. It is used as a weapon in cockfighting.

- **cornification**
  conversion of living cells lining tissues to a hard packed layer of dead cells.

- **coronary**
  applied to several structures in the body encircling an organ in the manner of a crown. The coronary arteries are the arteries of supply to the heart which arise from the aorta, just beyond the aortic valve; through them blood is delivered with great pressure to the heart muscle.

- **corpus luteum**
  active tissue which develops on the ovary at the site where an ovum has been shed. If conception is not occurred, the issue gradually disappears, if it is occurred the tissue becomes functional and produces progesterone.

- **correlation**
  a measure of how two traits

vary together. A correlation of +1.00 means, that as one trail increases the other also increases. A correlation of -1.00 means that as one trait increases the other decreases. A correlation of 0,00 means that as one trait increases, the other may increase or decrease. Thus, a correlation coefficient may lie between +1.00 and -1.00.

- **corrosive sublimate**
  a violent poisonous bichloride of mercury, highly effective as an antiseptic, germicide and disinfectant. As an antiseptic wound wash and in dressings, it is used in 1:1000 to 1: 5000 solutions; as disinfectant in 1:1000 to 1:5000 solutions.

- **cortex**
  the main shaft of the fibre, lying within a layer of cuticular scales and surrounding the medulla, if present, as in certain types of 'hairy' fibres. The cortex consists of long narrow cells.

- **cosmic rays**
  radiations of sun, mostly atomic nuclei (protons) with very high energies, originating outside the earth's atmosphere.

- **cosset**
  a lamb raised with no help of its dam.

- **cotted**
  thread which has become partially felted or matted whilst on the sheep's back.

- **cotted wool**
  wool that has felted together on the sheep's back.

- **cotton seed cake, cotton seed meal**
  an animal feeding-stuff made from the seeds of the cotton plant after the extraction of oil, mainly used for adult ruminants, has a high fibre content, is poor in calcium, with low protein quality. It is yellow-coloured due to the presence of gossypol which is ionic to young pigs and poultry and is available cither in decorticated or un-decorticated forms.

- **cotton seed hulls**
  the protective covering of cotton seeds in the outer layer.

- **cotton seed oil cake, decorticated**
  residue which remains after mechanical pressing of clean cotton seed, com-

posed, mainly of the kernel with such unavoidable portions of the hull and fibre as may be left in course of expression of oil.

- **cotton seed oil cake, undecorticated**
  the product obtained after mechanical pressing of clean, mature and unhulled seed from which maximum oil has been extracted.

- **cotyledon**
  area of foetal membranes which link the maternal caruncles of the uterus to form the placenta. Sometimes called as *buttons*.

- **cotyloid**
  cup-shaped.

- **coumarin**
  the compound which gives sweet clover its characteristic odour, a white, crystalline compound with a vanilla-like odour.

- **count**
  1. woollen yarn-the number of standard skeins, each 256 yards long, which weigh 1 pound.
  2. worsted yarn-the number of standard hanks of a worsted, each 560 yards long, which weigh 1 pound.
  3. raw wool-a number (quality dumber) used to indicate fineness.

- **counter**
  1. a general title applied to radiation-detection instruments
  2. survey meters that detect and measure radiation.

- **counter irritant**
  an agent causing a skin irritation which is intended to relieve another irritation of an organ below the skin.

- **countervailing duties**
  import assessments against government subsidised exports from other countries.

- **covariance**
  common variation between two traits due to same genetic and/or environmental effects acting on both traits.

- **cover**
  vegetative ground cover supplied by the above-ground portions or plants.

- **cover crop**
  crops planted between seasons for cultivated crops to protect the soil from erosion by wind or water, e.g. rye.

- **covert**
  a small feather paired with each primary feather in the wing.

- **covey**
  1. a flock of birds;
  2. quail, partridge.

- **cow**
  a mature female bovine which already had one or two pregnancies.

- **cow hocked**
  a condition of a cow or horse in which the hocks are close together and the fetlocks spaced out.

- **cow tail**
  hairy, coarse , tippy wool, usually from stronger crossbred types.

- **cowpox**
  an sensitive virus disease of cattle accompanied with a slight fever and a typical eruption which usually confined to the teats and udder of the female or the scrotum of the male. The affected animals are isolated on time as precaution to avoid carrying infection to other cows. Affected animals are to be bathed with cresylic disinfectant like a 3 % saponated cresol solution and antiseptic ointment or oil applied. Milking should be done gently and with as little discomfort to the animal as possible.

- **crab meal**
  the ground dried waste of the un-decomposed crab with the shell, viscera and part or all of the flesh. It must contain not less than 25% protein.

- **crabbing**
  the finishing process of passing the cloth over rollers into hot water or stern for preventing the formation of creases or irregular reduction.

- **crab-tree effect**
  the sustained development in the feed-use efficiency of an animal under feed restriction to reduce the feed supply repetitively in order to maintain a flat growth rate.

- **cracked wool**
  a fleece which felts together and cracks open, especially in wet weather.

- **cradle combed**
  a Wyandotte comb which is hollow in the centre.

- **craft**
  a set of work in agriculture such as machinery operation, milk production and maintenance, or plant nursery practice.

- **crammer**
  1. a machine for calorific poultry.
  2. an individual who fattens poultry.

- **cramp**
  an instinctive contraction of a muscular tissue in the body.

- **crampiness**
  a condition among cattle, especially bulls in confinement, in which instinctive contractions of muscles of the hind legs result in pain and discomfort.

- **cranial**
  pertaining to the skull or the interior or superior end of the body.

- **crawler**
  a newly hatched insect.

- **cream**
  an oily substance containing fats which rise on milk left to stand. In manufacturing, milk is centrifuged to divide it into cream and skim milk. The cream is sold as liquid, clotted or sterilised, or is churned into butter.

- **cream cheese**
  soft cheese made from cream trained through muslin.

- **cream separator**
  a machine which uses centrifugal action to divide milk into lighter cream for buttermilk and skimmed milk or separated milk containing the heavier particles.

- **creamery**
  a butter factory.

- **creamy plumage**
  the pale yellow pigment in newly grown feathers.

- **creatinine**
  a nitrogenous compound arose from protein metabolism and secreted with the urine.

- **creep**
  1. when wool fibres creep the move steadily slowly in one direction, causing felting.
  2. an area of access to piglets separate from the sow lying area.

■ **creep fed**
extra feed by means of small openings in panels that permit only smaller-sized animals (pigs, calves, or iambs) to enter.

■ **creep feeding**
a system of feeding young animals prior to weaning and designed to exclude mature animals.

■ **creep ovary**
ovary having a proportion of abnormal cysts.

■ **crenation**
recoil of a cell due to hypertonic solution. Liquid leaves the cell due to osmotic pressure.

■ **creolin**
an disinfectant, germicide, antiseptic and deodorant used to control certain external parasites of animals. It is a dark-brown liquid consisting of high boiling coal-tar, phenols and oils and forms milky emulsion when diluted with water.

■ **crepitations**
certain sounds occurred along with the normal sounds of breathing in various diseases of the lungs or bronchial tubes.

■ **crepitus**
the grating of fractured bones.

■ **crest**
the edge of an animal's neck.

■ **cribbing**
a condition in which the horse bite the manger or another object while sucking in air.

■ **crimp**
1. the waviness of wool.
2. One of the waves in wool fibres measured as the number of crimps per unit length and an important quality in assessing manufacturing value.

■ **criss cross breeding**
a system of pig breeding in which the same two breeds are used as alternative sires on home-produced, crossbred gilts.

■ **critical temperature**
fairly accurate temperature of the air when the heat lost by the animal body through physical means such as evaporation just balances the heat produced as a result of internal work and tissue oxidation and for which further tissue oxidation is

necessary to maintain the normal temperature of the body.

- **crock**
an old ewe or horse.

- **crooked foot**
a horse foot with one side wall higher than the other. It is largely attributed to inherited faulty conformation with 'toeing outs or 'toeing in'. It may also be caused by unequal pairing of the foot sole or by improper shoeing.

- **crop**
the food-collecting organ of an animal found in front of the chest.

- **cropping**
cutting or shearing the nap of fabrics to give a even face or a level nap.

- **cross bred**
the progeny from the random mating of pures or crosses. May refer to a first-cross.

- **cross bred wool**
wool from sheep produced by crossing any of the merino sub-breeds of sheep with any of the English long-wool breeds. The wool is supposed to be as coarse as or coarser than the huff-blood grade, or coarser than 64's on the spinning count scale

- **cross breeding**
the mating of animals (or plants) of different breeds but both of which are pure-bred for combining the best characteristics of the two breeds. The progeny known as cross breeds, possess hybrid vigour, often expressed in females as increased mothering ability. They are sometimes crossed with a male of a third breed to introduce further new breeds.

- **cross immunity**
immunity consequential of infection with one producing organism against another. For instance, rinderpest virus infection in dogs gives rise to a degree of immunity against canine distemper virus.

- **cross pregnancy**
growth of a foetus in the opposite horn of the uterus on which ovulation occurred.

- **crude fat**
Also referred to as *ether ex-*

*tract*, it is that part of a feed which is soluble in ether.

▪ **crushing mill**
a machine to crush cereal grains before feeding to livestock.

▪ **cryogenic flask**
a flask used in artificial insemination to carry semen at very low temperatures.

▪ **cryciscope**
instrument used to determine the freezing point of liquids.

▪ **cryotherapy**
the curative use of cold.

▪ **cryptorchidism**
a failure of the mammalian testicles to come down into the 'scrotum'.

▪ **crystal lattice**
the organised structure of a solid chemical element or compound the components of which are lightly packed together, such as lactose crystals.

▪ **cub**
1. a cattle pen or small enclosure such as a chicken coop or rabbit hutch
2. the young of certain animals e g. fox.
3. a crib or fodder rack.

▪ **cuber**
a machine which forces meal through circular holes to produce extruded pencils of meal which are chopped into cubes or pellets of various lengthy ranging from 3.25 mm for poultry to 13 mm for cattle. It is usually binding with molasses at the same time.

▪ **cuboidal**
shaped like a cube.

▪ **cuckoo breeding**
uneven barring where the two colours are indistinct and run into each other.

▪ **cud**
a bolus of previously eaten food which has been regurgitated by a ruminant animal for further chewing.

▪ **cull**
an animal below the standards of the animals in the remainder of the herd.

▪ **culling**
the procedure of eliminating non-productive or unwanted animals.

▪ **cultures**
a population of micro- organisms in a medium as in the in microbiology. A pure

culture contains only organisms that originally arose from a single cell. Cultures are used for manufacturing cultured dairy products and most cheeses.

- **cup comb**
  an atypical shaped comb as seen in the Sicilian buttercup.

- **curd**
  the gelatinous or thicken part of milk. Curd from whole milk consists of whey, fat and casein, whereas curd from skim milk contains casein and whey but only traces of fat.

- **cure**
  the preservation of meat by malting, smoking or pickling fur storage.

- **curry**
  to massage, clean and brush a horse, usually using a special comb called a curry comb.

- **curves**
  the sloped back and tail of a Malay and their relative angles.

- **cushion comb**
  another quaintly sniped comb as in the silkie.

- **custom freezing**
  the drawing and freezing of semen from a specified bull by someone who is not the owner of the bull on a contract basis.

- **customary host**
  one in which the parasite commonly matures and reproduces.

- **cuticle**
  1. the external surface of skin.
  2. the layer of flattened cells which protects the surface of the fibre.

- **cutter blower**
  forage blower.

- **cutting**
  1. the amputation of testicles by castration.
  2. the separation of one or more animals from a herd.

- **cyanacobalamin**
  same as vitamin $B_{12}$.

- **cyanosis**
  a condition of blueness appeared in the tongue, lips, mucous membranes of the eyes, etc., when there is a lack of oxygen in the tissues.

- **cyclic fluctuations**
  the oscillations in vacuum

level recorded where the king milk tube enters the claw piece.

- **cyclotron**

  a particle accelerator in which charged particles receive synchronised accelerations repeatedly with the help of electrical fields as the particles spiral outward from their source.

- **cytokinesis**

  1. division of the cytoplasm following the division of the nucleus in meiosis or mitosis cell division,
  2. the branch of biology which deals with the structure, function, properties, physiology, development and reproduction of cells.
  3. Disintegration or dissolution of cell, that pan of protoplasm, which externally, is in contact with the cell wall and the liquid with which protoplasm is saturated and internally with the plastids, nucleus, droplets and grannies suspended in it.

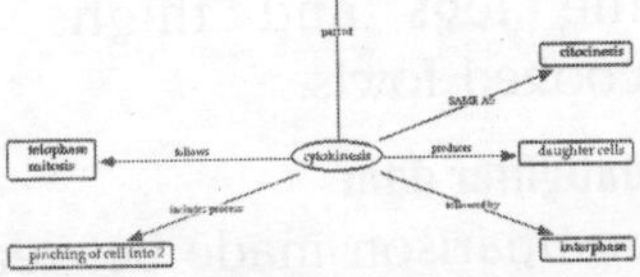


- **cytoplasmic inheritance**

  the transmission of hereditary characters via the DNA of self- replicating extra nuclear cells such as mitochondria and chloroplasts. Unlike the genes of the chromosomes, mutation in mitochondria and chloroplast DNA segregate in a non-Mendelian way, as the phenotypes of the progeny it determined by the number and phenotype of these organelles in the cytoplasm of the zygote.

- **cyloplasmic genetic male sterility**

  the type of male sterility which is dependent upon the action of genes carried in the nucleus with a particular cytoplasm.

- **cytoplasmic male sterility**

  type of male sterility which is governed by the cytoplasm rather than the action of nucleus. It is transmitted only through the female parent.

- **D value**

  the percentage of digestible organic matter in the dry matter of the feed.

- **dags**
staples or locks of wool on the breech area heavily coated with dung.

- **dairy character**
physical traits which suggest high milking ability. Important indications include an alert, long thin neck; mining head; openness and sharpness throughout; flat flinty bone and thin, pliable hide with a soft silky hair coat; prominent hip and pins; sharp withers: thin thighs.

- **dairy cows**
cows and heifers, as distinct from beef cows, kept for producing milk or for rearing calves for a dairy herd

- **dairy husbandry**
deals with the breeding, feeding milking of dairy cattle and the production and sale of milk.

- **dairy product**
any dairy production of milk like homogenised milk, cream , butter , buttermilk, cheese etc.

- **dairy records processing centres**
the computing facilities where information from the periodic tests in N.C.D.H.I.P. herds is summarised and analysed and where information to be used in future management decisions is produced and returned to the dairyman.

- **dairy temperament**
noticeable signs of health, conformation and quality which show the potential of milk which an animal will produce.

- **dandruff**
gathering of brain like scales on the skin. It becomes of exfoliation of the sebaceous glands

- **dapsone**
a bacteriostatic given by injection into the udder in cases of mastitis due to *streptococcus agalactiae* and by mouth in the treatment of *coccidiotis* in cattle.

- **dark chestnut**
a brownish black, mahogany, or liver-coloured horse.

- **dark meat**
the legs and thighs of cooked fowls.

- **daughter dam**
comparison made between

production records of daughters to their dams

- **daughter dam comparison**
method of estimating transmitting ability by expressing a cows yield as deviation from her dam's yield- this procedure has been superseded by several more accurate methods.

- **day old chick**
chick that is despatched to the buyer within 24 hours of hatching.

- **days dry**
the number of days dry prior to caking at the start of the production record listed.

- **deacon**
a veal calf that is marketed before it is a one week old.

- **dead wool**
1. wool collected in the paddock from sheep which have been dead some considerable time;
2. of inferior grade to plucked wool.

- **dead in shell**
1. a fertile egg in which the developing embryo had died.
2. embryo death occurring between the 14th and last day of incubation.

- **deadweight**
the weight of a dressed carcase.

- **deaeration**
removal of air. In the processing of milk, it refers to vacuum nation for removal of unwanted volatile substances that impart odours. No moisture is removed.

- **deaf ear**
an old name for the ear lobe.

- **dealer**
a person who buys and sells poultry of all ages.

- **deaminase**
an enzyme involved in the exclusion of an amino group from a molecule; ammonia is liberated in this process.

- **deamination**
exclusion of the amino group ($NH_2$) from an amino acid.

- **decalcification**
1. the removal of calcium from the bones of animal in negative calcium balance.
2. means the removal of calcium carbonate or calcium ions from soil by leaching,

a natural process in soil formation.

■ **decant**
to pour off the top layer of a multi-layered solution, or the liquid from a precipitate.

■ **decapitate**
1. to be head;
2. to cut off the head.

■ **decarboxylase**
an enzyme that liberates carbon dioxide from the carboxyl group of a molecule.

■ **decarboxylation**
the exclusion of a carboxyl group(— COOH) from an organic acid.

■ **decay**
the breakdown of organic tissue, usually associated with the presence of micro-organisms.

■ **decay (radioactive)**
the impulsive transformation o one-nuclide into a different nuclide or into a different energy state of the same nuclide. The process results in a decrease, with time of the number of the original radioactive atoms in a sample. It involves the emission from the nucleus of alpha particles, beta particles (or electrons), or gamma rays.

■ **deciduous teeth**
also known as milk teeth, it is the first teeth to appear in a young animal later shed and replaced by permanent teeth.

■ **deck feather**
either of the two central feathers in a bird's tail.

■ **decortication**
taking away of the bark, hull, husk, or shelf from a plant, seed, or root. Also, taking away of portions of the cortical substance of a structure or organ, as in the brain, kidney and lung.

■ **deep**
not near the surface, opposite of superficial.

■ **deep litter**
a system of bedding for poultry or cattle using sawdust or straw shavings.

■ **defecation**
the passage of waste material from the digestive tract of an animals.

■ **defect**
any blemish, fault, imperfection or departure from the standard.

- **defective unit**
unit that does not reach some prescribed standard.

- **deferred grazing**
withholding animals from a pasture beyond the normal start of the grazing season.

- **deficiency**
1. the loss of a terminal eccentric segment of a chromosome.
2. insufficiency in terms of amount, volume, proportion, a jack, a state of incompleteness.

- **deficiency disease**
a disease resulting from insufficient dietary ingestion of some nutrient.

- **definitive host**
(final host) the animal in which a parasite undergoes its adult and sexual life.

- **defluorinated**
has fluorine content reduced to a level which is non-toxic under normal condition.

- **defluorinated phosphate**
a useful mineral supplement which calcined, fused for precipitated rock phosphate from which nearly all fluorine has been removed.

- **deformation**
the twist of the whole body or any part of it which may be due to some accident or inherited one.

- **degeneration**
1. worsening or deterioration.
2. change of a tissue from a higher to a lower or less active functional form.

- **deglutition**
the act of swallowing.

- **degraded ecosystem**
in ecosystem of erratic communities resulted from a recent major disturbance and containing few topic levels. A graded ecosystem usually has a less complex food web than does a natural ecosystem. Sometimes, called a diselimex

- **degras**
the commercial name for crude grease of wool .

- **dehorn**
1. to take away the horns from sheep, cattle and goats
2. to treat young animals so that horns will not develop.

- **dehorning**
the amputation of an animal's horns as a measure

to prevent from wounding other animals. The best method is to disbud the young animal, but horns can also be amputated under anaesthetic by electric saw.

- **dehydrate**
  to take out most or all moisture, from a substance, primarily for the purpose of preservation.

- **dehydration**
  the removal of water from a substance.

- **dehydration synthesis**
  chemical union involving the removal of water.

- **dehydrogenase**
  an osidoreductase enzyme that catalyses the oxidation from a substrate. The removed hydrogen is donated to an acceptor other than molecular oxygen.

- **delaine wool**
  long stapled fine wool from the Middle Western states. It does not necessarily come from the delaine breed of sheep.

- **deletion**
  the non-appearance of a portion of a chromosome, causing genes to be absent or missing.

- **deltoid**
  1. having a triangular shape or outline. dementia.
  2. a general designation for mental deterioration.

- **demodex**
  a genus of mites. *Demodectic mange* in dogs is caused *Demodex canis*

- **demodicidosis**
  influx with demodectic mites. Abbreviated as dcmodicosis.

- **demography**
  the study of numbers of organisms in a population and their variation with time.

- **demonstration**
  presentation of the actual working of a thing or practice of the results of things actually done. The principle behind demonstration is that '*seeing is believing*'.

- **demulcent**
  a drug which gives a relaxing action particularly upon the mucous membrane or inflamed sting surfaces.

- **denaturation**
  an alteration in the original configuration of a rnacro-

molecule resulting in loss of its biological activity. Denaturation occurs from heat treatment, extreme pH changes, chemical treatment or other denaturing agents.

- **denature**

annihilation of the native stale of a substance.

- **dens**

1. a tooth.
2. a tooth-shaped projection.

- **desensitisation**

1. the process of reducing sensitivity.
2. an immunological transformation to a new slate of insensitivity.

- **density**

1. compactness in a fleece;
2. the close proximity of fibre growth on a giving surface of skin.

- **dental formula**

a formula indicating the number of each kind of teeth for a given animal species. The teeth are shown in the order incisors (i) canines (c). premolars (p) and molars) (m) with teeth numbers for one side of the upper jaw shown above those for the lower jaw.

- **dental pad**

the very concrete gums in the upper jaw of sheep, goats, cattle, and other ruminants.

- **dental pulp**

the spongy, jellylike central part of the tooth.

- **denticles**

projections, e.g., the denticles on the lips of large intestinal round worms enable them to injure the intestinal lining of their host.

- **dentition**

the physiology, type and number of the teeth of an animal. Animals having the same type of teeth are said to have homodont dentition whereas those whose teeth are not of the same are said to have *heterodont dentition*. The dentition of animals having two successions of teeth is known as *diphyodon't*; that of animals with only one set of teeth throughout life is *monophodont*. Animals whose teeth are continuously discarded and replaced throughout life have a *polyphyudont dentition.*

- **deodorant**
a substance which obliterates or masks bad odours, e.g. chlorine water, chlorinated lime, etc.

- **deodorisation**
the process of correcting, removing or masking undesirable odours. Deodorisation is one of the most important operations in fats processing; all edible fats are subjected to it before offering to consumers.

- **depigmentation**
the operation or result of removing pigment,

- **depilalion**
1. the process of the annihilation of hair that takes place during certain skin
2. chemical or thermal substances to the surface of the body.

- **depilatory**
a chemical preparation which removes hair from the skin, e.g., barium sulphate.

- **depot fat**
fat stored in the body.

- **depth of staple**
the length of staple on the body of the sheep.

- **derived proteins**
artificially synthesised protein-like compounds including compounds resulted from the decomposition of proteins.

- **dermatitis**
an inflammation of the skin.

- **dermato**
used as a prefix which means skin.

- **dermatoglyphics**
surface characteristics of the skin respect to its folds, furrows, ridges and winkles.

- **dermatology**
the science that relates to the skin in health and disease.

- **dermatophyte**
a fungus living as a parasite on hair, skin, claws or nails of man or animals.

- **dermatophytosis**
any skin infection caused by a fungus. e.g. ringworm.

- **dermatosis**
a pathologic condition of the skin.

- **dermatoltopic**
having a discriminating similarity for the skin.

- **dermis**
the inner layer of the skin laying immediately beneath the epidermis and consisting of a loose network of collagen and elastic fibres which support the principal blood vessels and nerve fibres.

- **dermoid cyst**
an encapsulated mass that contains ectodermic structures such as epithelium, teeth, glandular elements and hair in disorganised arrangement and incomplete formation.

- **desert**
1. place where rainfall is often less than 10 inches per year with extreme temperature variations. Most of the vegetation is restricted to living only during the very short rainy season.
2. a region with little rainfall/precipitation and, consequently, with scant vegetation, such as a steppe, tundra, dune and fell fled.

- **desiccated**
1. dried out;
2. exhausted of water or moisture content.

- **desiccation**
1. the process of drying,
2. to remove moisture from an object to a definite level.

- **design of experiments**
that part of the experiment which deals with the collection of observations.

- **desquamate**
the peeling of epithelial cells. To strip or come off in layers or scales as the epidermis in certain diseases.

- **detector**
a radiation-detection instrument.

- **detergent**
a compound which lowers surface tension used in mixed feeds because of apparent bacterio-static effects. It is also used for cleaning dairy equipment.

- **determiner**
same as gene.

- **detoxicants**
agents used to counteract the action of poisonous substances.

- **detritus food chain**
the food chain in which micro-organisms absorb and

break down the energy-rich compounds synthesised by he primary producers.

- **daughter herdmate deviation**
amount by which a cow differs in yield or other traits from others.

- **deutectomy**
the taking away of the yolk sac from newly hatched chicks.

- **developed acidity**
the amount of acids produced when micro-organisms ferment lactose in milk. It is usually expressed as a percentage of lactic acid, also called *real acidity* as opposed to *apparent acidity*.

- **development**
the ability of the merino for growing what is apparently more skin than is required to cover the body smoothly, with production of wrinkles or folds; varies in extent according to the type of merino,

- **deviation, standard**
the root mean square deviation of the observed values from their average.

- **devil's grip**
a serious defect in conformation, appearing as a depression immediately behind the fades and associated with yolk-stained, short wool of poor character. The condition sometimes extends down behind the shoulder blades add even right round the girth.

- **dew claw**
the rudimentary fifth digit protruding from the heel of dogs, pigs and cattle.

- **dewlap**
1. the fleshy wattle of the turkey.
2. the loose flesh which hangs from the throat of oxen and cows.

- **Deoxyribonucleic Acid (DNA)**
the chemical substance that carries the genetic material of the cells. The structure of DNA determining the structure of ribonucleic acid which in turn determines the structure of proteins of the cell. Its structure is of double helix.

- **dextran**
a plasma substitute, for use in transfusion instead of whole blood in cases of severe haemorrhage, etc.

- **dextrin**
an intermediate polysaccharide product obtained during starch hydrolysis.

- **DHIR**
Dairy Herd Improvement Registry, a dairy record-keeping plan sponsored by the breed associations.

- **dhrek**
an Asiatic tree of which the fruit and leaves are poisonous to farm animals.

- **diabatic**
a process that involves loss or gain of heat by the air from or to an outside source. It is also known as non-adiabatic process.

- **diabetes mellitus**
disorder in carbohydrate metabolism caused by a lack of insulin. characterised by excess glucose in blood and urine.

- **diagnosis**
the art of determining and distinguishing disease with the help of case history, laboratory techniques and symptoms.

- **diallel crossing**
type of crossing in which each of a number of males is made to cross to each of a number of females so as to compare the breeding value of males.

- **dialyser**
an equipment which could be separated a colloid from a crystalloid, the process is known as *dialysis.*

- **dialysis**
the parting of colloidal dispersed substances from truly dissolved crystalline substances in a given solution by passing them through certain membranes.

- **diamond**
a term applied to the wing bay or wing end.

- **diamond tup**
a shearing male lamb. Also called a din Mont, particularly in the borders.

- **diaphoretic**
an agent which increases the secretion of sweat, e.g., potassium nitrate.

- **diaphragm**
is the muscular and tedious structure which separates the chest from the abdominal cavity mammals.

- **diaphragm milk pump**
a milk pump where the milk

is moved by the to and fro action of a large diaphragm.

- **diaphysis**
  the streak of a long bone.

- **diarrhoea**
  waterly faeces which may be caused by internal parasites, bacterial infection, excessive feeding of green succulent or other feed. It is usually a symptom of disease. Simple diarrhoea could be treated with cathartics like caster oil. Extreme fatigue is one of its characteristics as the process of dehydration takes place during diarrhoea.

- **diastases**
  an older designation for the enzymes responsible for the breakdown of starches to sugars. The term is synonymous with enzymes in the French language since this change from starch to sugar was first recorded in the scientific literature by a French-man.

- **diathermy**
  a process by which electric currents can be passed into the deeper parts of the body so as to produce internal warmth and relieve pain, or by using powerful currents, to obliterate tumours and diseased parts bloodlessly.

- **diazinon**
  an anti-tick organo phiosphorus compound.

- **dicoumarol**
  a chemical compound originate in ruined clover and lespedeza hays. It is an anticoagulant and can cause internal haemorrhages when eaten by cattle. The popular name is dicumarol.

- **dicrotic**
  having two heart beat impulses. Pulse is one in which at each heart-beat two impulses are fell by the finger that is taking the pulse.

- **dielectric**
  conducting electricity poorly. Fat is such a substance.

- **diestrum**
  the part of a female's cycle between phases of estrus (heat),

- **diet**
  food or feed of an animal, which is having implications of amounts consumed, or offered, in any given time.

- **dietary diseases**
  diseases caused due to shortage of the animal needs for nutrients and vitamins.

- **dietectics**
  it deals with the application of the science of nutrition to the feeding of individuals and group of people.

- **differential(in milk pricing)**
  an amount, added to or subtracted from the price of milk at a specified base contain of fat or other components based on the level of fat or components in the milk.

- **differential segments**
  a block of genes in respect of which two pairing chromosomes differ, in a permanent hybrid, in contrast to a pairing segment where they pair and cross-over and are therefore homologous,

- **differentiation**
  1. the transformation of mother cells into different kinds of daughter cells (brain, kidney, liver. etc.). this process is irreversible.
  2. the process of acquiring individual characteristics as in the progressive diversification of cells and tissues of the embryo.

- **diffuse**
  not definitely limited or localised; widely distributed.

- **digestibility**
  that percentage of food taken into the digestive tract, which is absorbed into the body as opposed to that which is evacuated as faeces.

- **digestibility (indicator method)**
  a quick method of determining the digestibility of animal feed without measuring either the feed intake or faeces output. This is done by using an indicator with the feed to be tested and then determining the ratio of the concentration of the indicator to that of a given nutrient in the feed and also in the faeces resulting from feed.
  an ideal indicator has got the following qualities:
  1. should preferably a natural constituent of the feed to be tested. chromic for this purpose.
  2. totally indigestible,
  3. readily determined chemically,

4. un-absorbable from the digestive tract,
5. pass through the tract at a uniform rate,

- **Digestible Crude Protein (DCP)**
the most ordinary way of expressing the protein values and requirement for the ruminants in most of the countries. Digestible crude protein content of a food stuff is obtained by multiplying the crude protein content by the digestibility coefficient of the protein in the fodder.

- **digestible energy (de)**
the portion of the energy in a feed which can be digested or absorbed into the body by an animal.

- **digestible energy value**
the amount of utilisable energy in an animal food after deduction of the energy value of undigested faecal residues, (digestibility trial).

- **digestible nutrients**
that portion of the nutrients consumed which are digested and taken into the animal body. The term is generally applied to the proteins, carbohydrates and fats.

- **digestive protein (DP)**
the amount of protein of food that is absorbed by the digestive tract; it may be computed using the formula:
percent DP=percent crude protein of feed x digestion coefficient for protein in the feed.

- **digestion**
the transformation of food into living matter, i.e. the change of proteins, carbohydrates and fats, under the influence of enzymes, into substances which are assimilable.

- **digestive juices**
fluid materials secreted into the digestive tract. They are bile, saliva, pancreatic juice, gastric juices, juice, etc.

- **digestive tract**
the channel from the month to the anus through which food passes after consumption. It is subjected to various digestive processes. Primarily the stomach and intestines.

- **digital**
pertaining to the finger or the terminal part of a limb.

- **digitalis**
a valuable drug having di-

uretic properties, made from the dried leaves of the foxglove. It is used in the treatment of heart diseases.

- **dihybrid**
a double heterozygote such as AaBb.

- **dihybrid cross**
a cross involving two pairs of alleles, each of which regulates different characteristics.

- **diluent**
any substance used to make the semen less concentrated in artificial insemination.

- **diminishing return**
the point at which the expense of obtaining an increase in production becomes greater than the value of the increased production.

- **dingy**
applied wool lacking brightness, possibly the result of excessive yolk, unfavourable environment, or the presence of external parasites.

- **dioestrus**
the part of the female animal's cycle between periods of heat,

- **dip**
1. a dipping bath.
2. a proprietary chemical which is diluted in a recommended volume of water in a dipping bath for dipping animals, mainly sheep.

- **diplegia**
extensive paralysis on both sides of the body,

- **dipiold**
having two sets of homologous chromosomes, somatic cells are normally diploid, whereas gametic cells have only one set of chromosomes, i.e. are haploid.

- **diplotene**
the stage in meiosis when the chromatide become noticeable, and the homologous chromosomes begin to divide.

- **dipped sheep market**
a market, or a separate and distinct part of a market, at which, for a particular period, only dipped sheep are sold.

- **dipping**
consists in immersing animals in the insecticidal liquid.

■ **direct calorimetry**
measurement of the amount of heat produced within a small chamber (as a bomb cailorimeter, where the heat produced through combustion is absorbed by a known quantity of water in which the container is immersed. A means of determining caloric value of foods.

■ **direct expansion**
a refrigeration system in which liquid refrigerant is forced through a valve into a chamber of lower pressure. As it expends and becomes a gas, it takes on heat.

■ **direct comparison method**
process for estimating the relative genetic transmitting ability of a group of sires by comparing the relative yield of their daughters in the same herds and time periods.

■ **direct receipt**
the cash amount received from sale of livestock, crops and miscellaneous sources.

■ **dirt**
any foul substance, which at times cause dangerous animal diseases.

■ **dirt eating**
a habit of young animals of eating litter, soil, wool and other indigestible substances that may lead to the formation of balk of foreign material in the stomach. It may lead to death the animal due to stoppage of the exit from the stomach.

■ **disaccharide**
one of several so-called compound sugars which yield two monosaccharide molecules upon hydrolysis, sucrose, maltose and lactose are the most common.

■ **disbudding**
the removal of, or the prevention of the growth in the horn buds in calves.

■ **discontinuous distribution**
the occurrence of a species in one area, then in another area far from it, but not in any areas.

■ **discrete**
made up of disconnect parts, or characterised by lesions which do not become blended.

■ **discrete distribution**
method of depicting the fre-

quency of occurrence of different categories of traits which occur in discrete classes, such as sexes or months of calving.

- **disease**
any deviation from a normal state of health which temporarily impairs vital functions of animals.

- **disease cycle**
the chain of events taking place in the development of a disease.

- **dished**
a lobe which dips in the centre like a saucer.

- **dished bill**
the down curves in the upper bills of some ducks.

- **disinfect**
to destroy or render inert disease-producing germs (pathogens) and harmful micro-organism and to destroy parasites.

- **disinfection**
1. the cleaning of farm equipment and utensils (e.g. milking plants using suitable detergents and disinfecting agents (e.g.. sodium hypochlorite) to reduce bacterial contamination.
2. the cleansing of appropriate buildings, animals, implements and utensils. etc., which may be harbourin harmful bacteria or viruses after an infection has existed on a farm, the process can involve physical cleaning, the application of approved chemical solutions, steam cleaning and fumigation.

- **disinfestation**
external parasites. Mechanical or physical means of hindering their development, e.g. cleaning the yards of all refuse, removing litter and droppings frequently etc.

- **dislocating**
a method of killing poultry by pulling and twisting the fowl's head until the neck separates from the skull and the spinal curd is severed.

- **dislocation**
a displacement of a bone from its normal position in relation to a joint.

- **displacement activity**
an act performed by an animal that is apparently insignificant and immaterial to the situation with which it is faced. It usually occur

when there is blocking of the drive for the normal pattern of behaviour. The blocking of a drive may be caused by the presence of two conflicting drives.

- **display behaviour**
  actions performed by an animal that acts as specific signals, mainly to members of the same species. The activities may be movements, sounds, the sudden revealing of bright plumage etc.

- **disposition**
  the temperament or spirit of an animal.

- **dissect**
  to chop an animal into pieces for examination.

- **dissection**
  the chopped apart or cutting away of tissues.

- **dissipate**
  to cause to disappear; to spread in different directions.

- **distal**
  remote, as opposed to close or proximal.

- **distiller's dried grain with solubles**
  distillers grains always contain small amount of liquor after distilling off. When the material is spray dried to produce a light brown powder of variable composition it will be termed as *'distillers'* solubles'. These are often mixed with distillers' grains and dried together to yield a material sold as distillers' dried grains with solubles or dark grains.

- **distiller's dried grains**
  in the manufacture of alcohol and other distilled liquors from maize, rye or wheat the soluble materials are extracted leaving the residue. This residue after filtration is sold as wet or dried distiller's grains. The material is rich in crude protein (20-25%). lipid (5-8%) and crude fibre ('5-8%). in view of us fibre content the grain is seldom fed to swine or poultry.

- **distomiasis**
  means infestation with liver flukes

- **districhiasis**
  a double row of eyelashes.

- **diuresis**
  excessive discharge of urine.

- **diuretics**
  substances which produce diuresis, that is, which cause a copious excretion of urine.

- **diurnal variation**
  the amount of variation in one day.

- **dizygotic twins**
  a set of twins derived from two separate fertilised eggs. They are no more alike genetically than full brothers and sisters born separately.

- **dock**
  to cut off the tail (especially in sheep).

- **docking**
  to take away the whole or part of an animal's tail, lowland breeds of sheep are normally docked as lambs to keep away from the accumulation of dirt and faeces in the tail.

- **doddie**
  a polled cow.

- **doddle**
  also called tassel or wattle, it is loose flash hanging from the throat of a goat.

- **doe**
  an adult female deer, goal, or rabbit.

- **doeling**
  a female goat from its first to its second birthday, or from the first birthday until first parturition.

- **doggy**
  a term applied to short, coarse, glossy merino wool of non-descript type, frequently found on old sheep.

- **doggie**
  a motherless calf.

- **dolly**
  an ornament made of straw usually placed on top of a stack. Also called corn dolly.

- **DOM**
  Digestible Organic Matter, an assessment of nutritive value of food calculated from food dry matter content x per cent organic matter in food x digestibility of the organic mailer fraction.

- **domesticate**
  to bring a wild animal or fowl under control and in improve it through careful selection, mating and handling so that its products or services become more useful to man. Domestic animals breed under man's control.

- **dominance**
the tendency for one gene to exert its influence over its partner, after conception occurs and genes exist in pair. There are varying degrees of dominance, from partial co complete to over-dominance.

- **dominance hypothesis**
the hypothesis to explain the cause of hybrid vigour as 'the maximum number of dominant genes into hybrids'. It was proposed by Bruce in 1910.

- **dominance variance**
constituent of genetic variance which is due to dominance effect of genes.

- **dominant**
a gene that, when paired with its allele, covers up the phenotypic expression of that allele gene. For instance, black-and-white in Holsteins (B) is dominant over red-and-white (b). Thus. BB and Bb Holsteins are black-and-white, whereas bb individuals are red-and-white.

- **dominant (animal behaviour)**
specify the individual at the top of the social hierarchy. The dominant animal may behave aggressively to all other members of the social group and leads aggressive actions against attackers when the group is threatened.

- **dominant character**
one that is fully developed whether the gene producing it is present in one or both members of the chromosome pair, Ie,, whether the individual is heterozygous or homozygous.

- **dorking**
a table breed of fowl, square- bodied, variously coloured (including red, white, silver-grey and barred ). with a single comb and five-clawed white legs. Hens produce white eggs.

- **dormant**
resting condition for a living animal or plant with a relatively inactive metabolism.

- **dorsal**
pertaining to tie back surface of an animal.

- **dorset wedge silage**
silage created by filling a silo in wedge-shaped layers with cut grass or other

greenstuffs, covering with plastic sheeting and seating the silo to ferment.

- **dose syringes**

syringes having long dose pipes or nozzles and are used for careful oral administration of medicines to animals.

- **dose table**

drugs are administered according to label directions. assuming the dose for mature horse as 1.0, the dose for mature animals of other species is calculated as cattle 1.5, sheep and goals 0.s. swine u.i2. dogs 0.6 and cats 0.3. the dose is also influenced by the age of the animal.

- **duuble cross**

a blend of four inbred lines used in producing a hybrid chicken.

- **double fleece**

a fleece of wool belonging to the short wool group.

- **double laced**

two lacings of black as on an Indian game female's feather.. There is the outer black lacing round the edge of the feather and next the inner lacing.

- **double mating (double cover)**

1. mating a female livestock to two different males during the same heat period.
2. mating of female livestock with a male on successive days of an estrous period to enhance the probability of conception.

- **double pregnancy**

the existence of two sets of foetuses of different ages and born with a corresponding interval between litters, in the sow, cow, etc.

- **double standardisation**

the standardisation of milk for both fat and non-fat solids.

- **double suckling**

a system of feeding beef cattle in which a second calf is introduced and allowed to suckle alongside a cow's own newborn calf. The second calf is sometime rubbed with part of the placenta to encourage the cow to accept it as her own. The system is only used with cow producing sufficient milk to feed two calves.

- **doubles**

the term refers to twins, especially in case of lambs.

- **down**
  early hairy covering of baby chicks, ducklings, etc. Also the feathery part of the feather below the web and small tufts sometimes seen as faults on toes and legs of clean legged breeds.

- **down breeds**
  hornless, short-woollen breed of sheep, with coloured faces and legs, giving good dense fleeces and high quality meat.
- **down calver**
  a heifer or cow about ready to give birth to a calf.
- **down wool**
  wool from the breeds of sheep that originated on the downs of England.
- **dowse**
  chaff.
- **draft animal**
  an animal, e.g. a horse, mule or ox, used to draw heavy loads.
- **drafted animal**
  an animal either added to or separated from a herd or flock. Draft ewes are those sold from a breeding flock of sheep whilst they stiff are able to produce lambs. usually they are sold from a hill flock and sent to a lowland farm when they arc considered to lack hardiness for continued hill breeding) normally after producing three crops of lambs,
- **drain cocks**
  a manual or automatic cock placed at the low points in a line to drain off accumulated fluids.
- **draining pen**
  a pen into winch sheep emerging from a dipping bath are directed to allow surplus dip to drain off the fleece and return via a sieving device to the bath.
- **drake**
  the male of the duck family.
- **drakerel**
  a young drake.
- **drape**
  a cow or ewe unable of bearing offspring.

**draught horse**
a horse used to draw carts, trailers and farm implements e.g., the shire.

**drawing**
the various processes, including gelling, reducing and roving by which top silvers are converted into roving of the required thickness for spinning.

**drench**
the oral administration of medicine in the form of a drink through mouth or nose and forced down the throat of the animal.

**dressed fowl**
a dead fowl cleaned and plucked ready for cooking.

**dressing**
any germ-free material applied to a wound for its protection and medicated bandages.

**dried blood**
a dried organic fertiliser derived from waste blood from slaughterhouse. It has a nitrogen content of about 13% and was dried by evaporation.

**dried grass**
grass, which has been dried by artificial method after cutting and is used as an ingredient in animal foodstuffs. It is either milled and mixed with molasses to produce cubes or pellets, or pressed directly into cubs or waters. It has been using for many years for cow, pigs and poultry becoming popular as a dairy cow ration combined with a concentrate and roughage. sometimes called dehydrated fodder.

**duct**
a tube in the body for the passage of liquids including secretions of glands, e.g., ancinus, oviduct, ductless glands.

**ductless glands**
glands which does not have discharging ducts. Their secretions are poured directly into the blood.

**dug**
a nipple or udder, chiefly of a cow

**dummy**
an equine that survives an attack of sleeping sickness but retains evidence of permanent injury to the brain

or spinal cord.

- **dump box**
  a big portable storage place similar to a forage wagon into which trailer loads of forage from the field are tipped and from which the forage is delivered in an even steady flow, usually by a moving floor mechanism or elevator belt, for loading into the silo. It may also be used lo deliver forage directly into a manger.

- **dump rake**
  an expression for a hay rake.

- **dumped wool**
  a package of wool compressed tightly for shipment and held by metal bands or wires.

- **dung**
  the faeces of animals often spread on the land.

- **dung liquor**
  the dark coloured liquid which drains from a manure heap, containing soluble compounds of ammonia and organic matter.

- **duodenum**
  the upper portion of the small intestine which extends from the stomach to the jejunum.

- **duplication**
  the process in which a chromosome is attached to a portion of its own homologous chromosome, thereby giving duplicate genes.

- **dyad**
  two sister chromatids :hat are linked together (synapsed) giving a tetrad.

- **dye reduction test**
  test of measurement of activity of bacteria in a sample of milk measurement of rate of consumption of oxygen in milk due to bacterial growth and activity. Indicators like methylene blue or resazurin are used to decolourise milk when oxygen level in it falls. Time taken for decolourisation provides indication of bacterial numbers and keeping quality of milk.

- **dys**
  is a prefix meaning painful or difficult.

- **dysentery**
  is a condition in which blood is discharged from the bowels with or without diarrhoea.

- **dysfunction**
partial abnormality, disturbance or impairment in the function of an organ or system.

- **dysgenesis**
a defect in breeding so that hybrids cannot mate between themselves, but may produce offspring with members of either families of their parents. Such offspring are sterile.

- **dyspepsia**
the impairment of the function of

- **dysphagia**
difficulty in swallowing.

- **dysplasia**
non-appearance of some part of the body.

- **dyspnea**
difficult breathing—laboured breathing.

- **dystocia**
irregular or not easy labour, causing difficulty in delivering the foetus and placenta.

- **dysuria**
an absence of urine.

- **eanling**
a young lamb.

- **ear lobes**
the small un-feathered area of skin just below the ears.

- **ear mark**
a distinctive mark clipped out of the ear of a sheep.

- **ear notcher**
a punch used for cutting botches in the cars of animal for identification.

- **early feathering**
an inherited tendency for early development of mature type of feather.

- **early maturity**
the completion in an animal of development of form and function sooner than others of the species.

- **early maturing**
reaching of sexual development at an early age.

- **ease**
to milk a cow a little for relieving pressure on its udder.

- **easy feed**
a method of feeding livestock in which the stocks are allowed easy access to feed. The feed is transferred mechanically to large hoppers or feeding passages.

- **eatage**
1. grass used for grazing as distinct from for hay.

2. the right to use aftermath for grazing.

- **ecbolics**
  drugs that cause contraction of the muscle fibres of the uterus, such as ergot, pituitrin, etc., used in cases of inertia at parturition.

- **ecchymosis**
  the collection below the skin of blood which has escaped from the vessels in the neighbourhood, as to a bruise,

- **eccrine**
  glands or tissues that secrete a substance without a breakdown of their own cells, e.g., sweat glands.

- **eclampsia**
  a disease occurred after parturition or during the later stages of pregnancy and characterised by loss of consciousness or convulsions, or both.

- **eco-climate**
  the climatic environment of a living organism.

- **eraseur**
  a surgical instrument used for castration of the larger domestic animals.

- **ecology**
  the study of living organisms in relation to their environment which embraces all communities of plants and animals interact with one another and with their physical environment.

- **ecosystem**
  1. system of an area including the plant, animal and other environmental factors land, water, micro-organisms, etc..
  2. all the factors of its non-living environment and the all inclusive term characterising a livin; community.

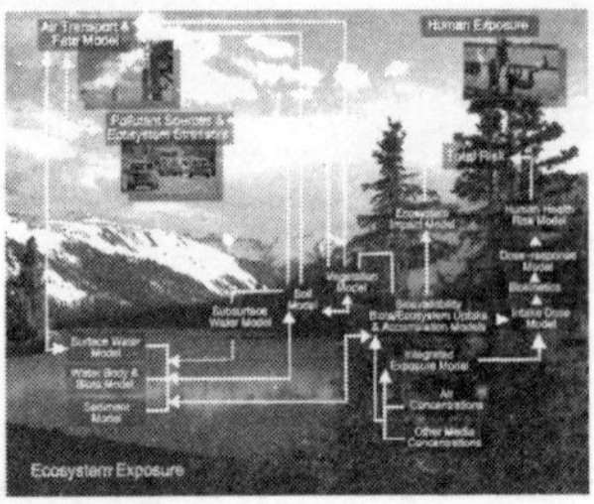


- **ecotone**
  the transition zone between two adjoining communities.

- **ecto**
  a prefix which means on the outside.

■ **ecto-commensal**
an organism that survives on the outer surface of another organism, the host, without either benefiting or injuring it.

■ **ectopic**
out of the usual place. Ectopic pregnancy is one in which a foetus is present outside the uterus.

■ **ectropion**
is a condition of the eyelids, in which the skin is so contracted as to turn the mucous membrane lining of the lid to the outsider ecotype
1. a biotype- resulting from selection with reference to a particular habitat.
2. an element of a linnaean species, separable as being associated with particular habitat factors.

■ **ectoblast**
an embryonic cell layer.

■ **ectoparasite**
a parasite that survives on the outer the body of its host,

■ **ectothrix**
forming a covering of fungus spores on the outer part of a hair.

■ **ectozoon**
outer animal parasite.

■ **eczema**
an inflammatory condition of the skin. It is not a contagious disease. It is usually associated with digestive derangement, improper feeding and unsanitary environment, affected animal exhibit an etching of the ears or legs and soon raw. Bleeding areas appear, treatment involves washing the affected area with a fairly strong saponated cresol solution, allowing it to dry and then applying an antiseptic dusting powder.

■ **edema**
the presence of abnormally large amounts of fluid in the intercellular issue spaces of the body, as in swelling of mammary glands commonly accompanying parturition in many farm mammals.

■ **edge**
an edge is formed by intersection of any two adjacent faces.

■ **effective half-life**
half-life of a radioactive decay and biological elimination.

- **effective rainfall**
  precipitation falling during the growing period of the crop that is available to meet the evapo-transportation requirements of crops. It does not include precipitation lost through deep percolation below the root zone or through surface runoff.

- **effecter**
  a cell or organ that is specialised to perform an action relative to the environment in response to a stimulus from the nervous system.

- **effectors**
  include muscles, cilia and other motile systems, glands etc.

- **effeminate**
  a wool that is thin and lacking in bulk or substance.

- **efferent**
  1. leading away from a centre;
  2. opposed to afferent.

- **efficacy**
  effectiveness, for instance, the effectiveness of an antibiotic in the therapy of a certain disease.

- **egg bound**
  a term used for a bird unable to lay an egg.

- **egg case**
  a protective layer or case for transport of eggs.

- **egg cell**
  the laying of eggs by hen incessantly without a break of continuity.

- **egg classes**
  three quality classes of eggs are: Class A—'fresh eggs'; Class B— 'second quality or preserved eggs'; and Class c —'non-graded-eggs' intended for the manufacture of foodstuffs for human consumption.

- **egg membrane**
  the layer of tough tissue coating an egg shell.

- **egg tooth**
  tiny horny projection on tip of a chick's beak, used to chip egg shell when hatching it drops off soon after the chick is out of the shell.

- **egg weight grades**
  there are seven weight grades for Class A and B eggs. Grade 1 are 70 g or over, Grade 7 are under 45g.

- **egg white**
  the albumen of the egg.
- **egg yolk**
  the central yellow portion of a hen's egg. It contains the egg cell.
- **eggery**
  a place where eggs are laid.
- **eggler**
  a dealer of eggs.
- **ejaculate**
  the quantity of seminal fluid (sperm) which comes out of a penis at the time of mating.
- **ejaculation**
  sudden or rhythmic discharge of sperm and seminal fluid from the male ejaculatory duct. The duct carries semen to the urethra, specially in mammals.
- **elastic fibres**
  stretchable fibres consisting of elastin fibrils held together by a cementing substance and entwined among collagen bundles of the dermis.
- **elasticity**
  the power of wool fibres to return to their original form after being stretched or compressed.
- **elasticity of demand**
  the tendency of demand for a commodity to be influenced or changed by various factors. 'pre-elasticity' is the tendency of demand to change as price goes up or down. 'income elasticity' is the tendency of demand to change as the consumer income goes up or down. both are related to ability to buy,
- **elastin**
  fibrous, yellow, connective tissue in skin, ligaments, tendons, blood vessels and all other supporting and protective tissues.
- **elastrator**
  an instrument used to stretch a strong rubber ring so that it may be placed over the neck of the scrotum for the purpose of castration.
- **electric dog**
  a mechanism to encourage cows to enter a milking parlour from a collecting yard, comprising an electric wire sandwiched between two parallel side wires which is drawn up behind the cows gradually as they go

through the parlour, by means of a cord in he parlour (electric fence).

- **electric fence**

a system of mobile fencing usually used to facilitate controlled grazing, consisting of easily moved insulated posts, usually about 3 ft. high, supporting thin wires (usually one for cattle and horses, two for sheep and pi) carrying an electric current, either battery generated or from the mains. The current is not continuous but in 'pulses' at the rate of about 60 per minute.

- **electro cautery**

it is useful for operations where space is restricted, such as removing small tumours, etc. in mouth, nose, or throat and to check haemorrhage in the deeper parts of wounds and sometimes fur disbudding.

- **electrocardiogram (ECG)**

a record of the alterations in electric potential that accompany each cycle of cardiac activity such alterations arc measured at the surface of the body, usually by electrodes attached to the arms and legs and are recorded as a series of waves through an oscilloscope or similar device.

- **electrolyte**

any solution that conducts electricity by means of its ions.

- **electromagnetic radiation**

radiation consisting of electric and magnetic waves that travel at the speed of light (e. g. x-rays, gamma rays , radio waves, light waves.).

- **electromagnetic spectrum (ES)**

a series of waves of varying length. At one end of the ES are gamma rays, whose waves measure only about one-billionth of an inch and at the other end are radio waves, which may be many miles length between this extremes, in order of increasing wavelength, are x-rays, ultraviolet light, visible light, infrared light, micro-waves and shorter radio waves. All waves have different characteristics, but all travel at the same speed (186,000 miles per second).

- **electron (symbol e)**

an elementary particle with a unit negative electric

charge. Electrons surround the positively charged nucleus and determine the chemical properties of an atom.

- **element**
  one of the 103 known chemical substances that cannot be separated into simpler substances by chemical means. A substance the atoms of which all have the same atomic number, e.g., hydrogen, lead, uranium, etc.

- **elevator**
  a machine for raising a cut crop, grain, or hales to the top of a silo, storage bin, or bale stack, etc., usually for storage, various designs exist. grain elevators usually consist of a continuously moving belt bearing buckets, or an endless chain bearing flights 01 steps (operating like a moving staircase elevators for hay or straw bales often comprise a continuous moving chain bearing spikes or cross slats.

- **elute**
  to wash off, or remote from; specially to take away by extraction, chemical compound from a substance on which it has been adsorbed.

- **emaciated**
  an excessively thin condition of the body.

- **emaciation**
  the loss of flesh causing extreme leanness.

- **emasculation**
  excision of the penis. castration

- **embden**
  a weighty white breed of goose with blue eyes.

- **embolism**
  the blocking of a small blood-vessel by a piece of some material that has been carried into it from larger vessels. It is generally due to fragments of a blood-clot which has formed in some other part of the body and have been carded by the circulation.

- **embryo**
  phase in the growth of an individual between conception and the completion of organ formation.

- **embryo mortality rate**
  percentage of dying fertilised ova while in the embryonic stage.

- **embryo stage**
  of the embryonic offspring up to 4 weeks post coitum.

- **embryo transfer**
  a process which involves the taking away of a developing embryo from a mother's uterus at a very early stage and transfer into the uterus of a recipient female.

- **embryology**
  the science relating to configuration and growth of the embryo.

- **emissivity or absorptivity**
  ratio of the rate of emission (or absorption) of radiant energy per unit area by the given body to the rate of emission (or absorption) from a blackbody at the same temperature.

- **emitic**
  an agent which causes vomiting, e.g., copper sulphate.

- **emollients**
  substances that have a softening and relaxing effect upon the tissues of the body to which they are applied. They are chiefly used upon hard dry skin. The chief are ointments, glycerine, soap, lard oils and fats.

- **emotion**
  a complex state of an individual in which certain ideas, feelings and usually motor expressions combine to produce a condition recognisable as such by the individual and frequently by others

- **emphysema**
  an irregular presence of air in some part of the body.

- **empty days**
  the number of days per reproductive cycle that a sow spends neither pregnant nor lactating.

- **empyema**
  the presence of pus in any body cavity.

- **emulsify**
  to disperse small drops of one liquid into another liquid.

- **emulsion**
  1. a preparation of one liquid distributed in small globules throughout the body of a second liquid.
  2. a milk-like product.

- **enamel**
  the white calcified tissue that covers the exposed surface (crown) of a tooth.

- **enchondroma**
a tumour formed of cartilages.

- **encephalitis**
an inflammation of the brain causing various central nervous system disorders.

- **encephalomalacia**
a condition characterised by softening of the brain. a disease causing lesions of the brain in young poultry, caused by a deficiency of vitamin E in the diet.

- **encysted**
parasites, collections of pus, fluid, etc, which are shut off from the surrounding structures by a ring of dense fibrous issue or by a membrane.

- **endarteritis**
inflammation of the inner coat of an artery.

- **endemic**
1. the normal population level of a potentially destructive species.
2. confined in its indigenous occurrence to a particular region.

- **endemism**
the situation of organisms being restricted to a particular region, such as an island.

- **endobiotic**
an organism living within its substratum.

- **endocarditis**
inflammation of the smooth membrane that lines the inside of the heart.

- **endocrine**
pertaining to glands that produce secretions that pass directly into the blood or lymph instead of into a duct (secreting internally). Hormones are secreted by endocrine glands.

- **endocrinology**
the field of study of the structure and function of the endocrine glands and the role of their secretions (hormones) in the chemical integration of the organism.

- **endoderm**
inmost of the three germ layers of the embryo.

- **endoenzyme**
an enzyme formed within the cell and not secreted into the medium.

- **endogenous**
internally produced inside the body, e.g., hormones

and enzymes.

- **endometritis**
an inflammation of the mucous membrane that lines the womb or uterus.

- **endometrium**
the mucous membrane that coats the uterus.

- **endoparasites**
parasites that live in the body tissues or cavities e.g., the tapeworm.

- **endophily**
the feeding or resting of insects in houses (inside).

- **endoplasmic reticulum**
the sub-microscopic, double-layered membrane, about 100 A° thick, consisting of canali-culi and cisterna. They are much folded in the cytoplasm and have microsomal particle attached to them, these are usually ribosomes. This association is responsible for the major biosynthesis of the cell.

- **endosteum**
a sheath of fibrous tissue lining the marrow cavity of a bone.

- **endosymbiosis**
a relationship of mutual benefit to two organisms.

- **endothelium**
layer of epithelial cells lining the cavities of heart, blood and lump vessels and the serious body cavities.

- **endothrix**
development within the hair shaft without a conspicuous external sheath of spores.

- **endotoxins**
toxic substances causing typhoid retained inside the bacterial cells until the cells disintegrate.

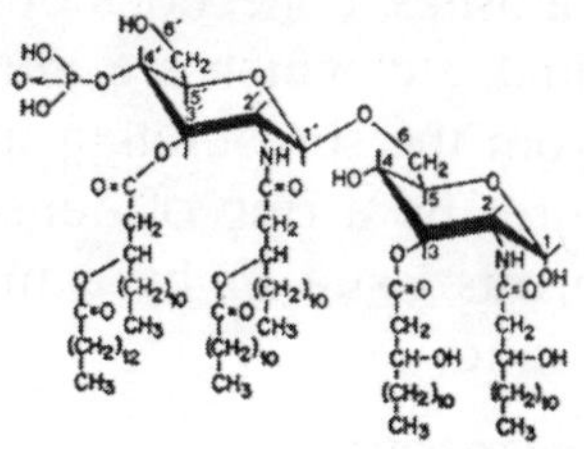

- **ENE**
Estimated Net Energy.

- **energy**
the capacity to perform work.

- **energy value**
the amount of heat each gram gives out on complete combustion.

- **ensilage**
a green chop (forage) pre-

served by fermentation in a silo pit or stack, usually in chopped form. It is also known as *silage*.

- **ensiled**
  having been subjected to anaerobic fermentation to form silage.

- **ensiling**
  the process of making ensilage (silage).

- **enteric**
  pertaining to tissues which line the intestine.

- **entuzoa**
  internal animal parasites, e.g., stomach worms.

- **entrails**
  the inner organs of animals, specifically the intestines.

- **enteroliths**
  posture that develop in the intestines, being formed by deposition of sails round a bard metallic or other nucleus.

- **entiopion**
  turning in the eyelids.

- **envenomisation**
  the injection of a poison into an animal.

- **environment**
  the sum total of all external condition, which affect the life and act of an animal.

- **environmental**
  pertaining to surrounding influences.

- **environmental variance**
  the part of the phenotypic variance which is caused by environmental effects.

- **enzootic abortion**
  a virus infection of sheep affecting the placenta, causing abortion about 10-14 days prior to lambing time. Common in South-East Scotland and North-East England.

- **enzootic disease**
  an animal disease prevalent in particular areas or district or during certain seasons (e.g., enzootic abortion) as distinct from epizootic diseases which spread rapidly and attack animals in large numbers almost simultaneously over large areas, even over entire continents ( foot and mouth disease).

- **enzootic hepatitis**
  an acute viral infectious disease affecting cattle, goat, sheep and man, it is characterised by transient fever, anorexia, profuse saliva-

tion, abdominal pain, diarrhoea and abortion in pregnant animals, it is caused by a pantropic virus with a pronounced hepatotropic attribute.

- **enzymatic**
  related to an enzyme.
- **enzyme**
  an organic catalyst produced within an organism. an enzyme increase the rate of reaction without altering the equilibrium constant for that reaction. Enzymes are highly specific in respect of the nature of the reaction catalysed and of the substrates utilised.
- **eosinophil**
  type of white blood cell that has a particular similarity fur the dye eosin.
- **epharmonic**
  differing from the normal or usual because of environmental influences.
- **ephemer**
  designating a very short-lived animal.
- **epicondyle**
  a bony prominence located above a condyle.
- **epidemic**
  the outbreak of an epizootic disease affecting great numbers of animals in one area at the same lime and transmitted from place lo place. the disease may riot necessarily present in an area.
- **epidemic tremor**
  an infectious viral disease which mainly affects young chicken. The disease has been characterised by ataxia, muscular tremors and paralysis.
- **epidemiology**
  science dealing with development and/or decline of an epidemic wi(h special reference to the relationships of various factors that determine the frequencies and distribution of infectious diseases.
- **epidermis**
  the external non-vascular layer of the skin, consisting of a horny layer or cuticle, an intermediate layer of active cells and a foundation layer from which the hair roots generally start their development. Also called skin.
- **epidermoid cyst**
  a keratinous mass in the skin enclosed by an epithe-

lial lining. The cells of sebaceous glands seem to be the origin of these cysts.

- **epididymis**
 a small, tortuous tube leading from the testicle. a site of sperm storage and maturation.

- **epigastrium**
 the area lying in the middle of the abdomen immediately over the stomach.

- **epigenesis**
 a hypothesis of embryogenesis, which states that development consists in the successive formation of new parts that do nut pre-exist in the fertilised embryo.

- **epilepsy**
 a chronic nervous disorder characterised by a complete loss of consciousness in a sudden and is associated with muscular convulsions.

- **epinephrine**
 1. a drug used to stop haemorrhage and to stimulate heart action.
 2. a slaughter house by-product obtained from the adrenal glands.

- **epiphyseal**
 pertaining to an epiphysis (a portion of bone attached for a time to a bone by cartilage, but subsequently becoming consolidated with principal bone).

- **equine**
 the horse and its relatives, the mule and the ass.

- **equine influenza**
 an equine disease caused by filterable virus. The abortion syndrome is characterised by high rate of abortion in pregnant mares, while respiratory syndrome is characterised by feverish reaction, nasal catarrh, depression and anorexia

- **equine malarial fever**
 an infectious viral equines disease characterised by a uneven course and is usually associated with intermittent fever, progressive weakness, oedema and anaemia of a progressive or transitory type. It is caused by a filterable virus.

- **equine plague**
 African horse sickness.

- **equine viral rhinopneumonitis**
 equine influenza.

- **eradication area**
 a region in which a programme designed to

eradicate a particular animal disease is in operation. It often involves the compulsory slaughter of animals or herds found to have the specific disease.

- **ergosterol**

a plant sterol that, when activated by ultraviolet rays becomes vitamin $D_2$. Also called *pro-vitamin D2* and *ergosterin*.

- **ergot**

1. a small piece of horn, hidden by a tuft of hair and located on the back of a horse's fetlock
2. a fungal disease of grasses, mainly cereals and rye. It attacks the seed heads and destroys the grain. Animals can suffer serious poisoning (ergotism) by eating diseased grain or products derived from such grain (e.g., bread, maize meal). In human beings it causes a situation known as St. Anthony's fire.

- **eriology**

a field of study of the theory of the cause of diseases.

- **erosion (animals)**

the gradual destruction of body tissues due to pressure, inflammation irritation etc.

- **error**

1. in the theory of testing of hypothesis, the probability of rejecting a hypothesis when it is actually true, is known as type I error. In quality control, it is equivalent to the producer's risk.
2. the difference between an observed or computed value of a quantity and the ideal or true value of that quantity.

- **error, experimental**

the variation between plots or other experimental units, due to causes other than the treatments that have been applied. may refer to a variance or a standard error. Does not imply a mi slake, but limitations of material and method.

- **error means square**

the residual or error of squares divided by number of degrees of freedom on which the bum is error of first kind. if, as a result of a statistical test, a statistical hypothesis is rejected when it ought to be accepted, i.e.. when it is true, then error is committed.

- **error of second kind**
  if as a result of a test, a statistical hypothesis is accepted when it is false i.e.. when it should have been rejected, then an error is made.

- **error (SS)**
  a measure of the dispersion of the varieties within the sample about their respective sample means.

- **error standard (SE)**
  a measure of the extent to which repealed determinations of a quantity, of a sample, mean, or of a comparison between the means, would vary among themselves, the standard error can be obtained by dividing the standard deviation by the square root of the number of observations.

- **eructation**
  the act of casting or belching up gas from the stomach.

- **eruption**
  a fast developing lesion of the skin due to and marked by redness, prominence or both.

- **erysipelas**
  an infectious disease, mainly of pigs, in which hot reddish inflammations of the skin and high fever occurs and severe lameness and difficult breathing follows. The disease commonly causes infertility and abortion.

- **erythema**
  reddening of the skin of swine which occurs due to congestion of the blood capillaries near the surface-application of cold water or alcohol gives relief.

- **erythrocyte**
  red blood cells which carry oxygen.

- **eryrhropoiesia**
  the production of erythiocytes (the red blood cells that transport oxygen). Occurs in the red bone marrow.

- **escherichia coli**
  a bacterium found in the alimentary canal of most mammals. certain stains cause disease, e g., *Coli Seplicaemia* in poultry and diarrhoea in various animals. The abbreviated form is *E. Coli*.

- **escutcheon**
  the part of a cow which extends upward just above

and back of the udder, where the hair turns upward in contrast to its normal downward direction. Also known as the milk mirrow.

- **esophagus**
the passage way leading from the mouth to the stomach. sometimes called the gullet.

- **essential amino acid**
an amino acid that cannot be synthesised in the body from other amino acids or substances or that cannot be made in sufficient quantities for the body's needs.

- **essential amino acids index**
method used for measuring protein quality to which all the ten essential amino acids are considered- it may be defined as the geometric mean of the egg ratios of these acids to the food acids.

- **essential elements**
1. any element necessary for the growth and development of an animal or plant. 2. a critical element, trace element.

- **estrous cycle**
cycle of events from one 'heat period' to the next.

- **estrus (oestrus)**
the repeated, restricted period of sexual accessibility (heat) in females marked by deep sexual urge. Also called estrus period.

- **ether extract**
one of the fractions determined by the proximate analysis of animal feeding-stuffs, containing mainly fats and oils together with various minor compounds, e.g. fat soluble vitamins.

- **ethmoid**
a bone which divides the nasal cavity from that of the brain.

- **ethology**
the science of animal behaviour.

- **ethylene**
a colourless inflammable gas sometimes used as an anaesthetic for small animals.

- **etiologic**
relating to aetiology.

- **aetiology**
the field of study of the causes of disease, both. direct and predisposing and the mode of their operation.

- **eucaryotic cell**
a type of cell which possess well defined nuclear membrane.

- **euchromatin**
non-condensed, active chromosomes or chromosome regions of the interphase nucleus.

- **eugenic**
that which tends to increase the fitness of the race.

- **eukaryote**
cell type of plants and animals other than the bacteria and blue-green algae; such a cell has a nucleus surrounded by a nuclear membrane.

- **euploid**
the occurrence of chromosome numbers in the-individual which is a whole - number multiple of the basic or haploid number.

- **eupnea**
ordinary quiet breathing.

- **european breeds**
those, such as charolais, simmental and limousin, native to Great Britain or continental Europe.

- **eutrophic**
a descriptive term for nutrient or soil solution with nutrient concentrations optima t or nearly optimal for animal or plant growth.

- **evaluation**
process of determining potential of a collection of accession.

- **evaporated milk**
the liquid food obtained by partial taking away of water from milk. The MF and TMS contents are not less than 7.5 and 25.5 per cent respectively. It contains 25 IU vitamin D per ounce and is homogenised. The food is sealed in a container and is so processed by heat either before or after sealing, as to prevent spoilage.

- **evaporation (transpiration)**
as related to environmental physiology, the loss of moisture (sweating).

- **evenness**
regularity of the fleece, mainly in terms of quality number and length.

- **eversion**
turning the inside out, applied to the change in position of the oviduct, at the time of mating and of artificial insemination.

- **eviscerate**
  to remove the entrails, lungs, heart, etc., from an animal or fowl when preparing the carcass for human consumption.

- **eviscerated**
  having had the viscera or bowels removed.

- **evisceration**
  of poultry and fish, refers to removal of the intestinal tract and other organs from the carcasses.

- **excise**
  to cut out or off.

- **excoriation**
  a superficial loss of epidermis caused by scratching, biting or other physical damage.

- **excreta**
  the waste products of the body which are eliminated through the excretory organs of the animal.

- **excrete**
  to cast off products from the cell.

- **excretion**
  removal of waste products from the body.

- **excystation**
  the escape of a cyst; especially, a stage in the life-cycle of parasites occurring after the cystic form has been swallowed by the host.

- **exergonic**
  energy yielding as in any chemical reaction which is accompanied by liberation of energy.

- **exfoliate**
  to remove the surface scale or layer.

- **exfoliation**
  the separation in layers or scales or dead bone or of skin.

- **exocrine**
  pertaining to secretions elaborated into a duct and conveyed to other organs or to the outside of the body.

- **exogamy**
  out-breeding.

- **exogenous**
  applied to an animal as -.m artificial additive to its own body resources.

- **exomphalos**
  a hernia which has escaped through the umbilicus (navel).

- **exophily**
  the feeding or resting of in-

sects out-of-doors.

- **ekoskeleton**
  the skeleton that lies outside the body tissues of an animal. It protects and supports the internal organs and may provide attachment for muscules.

- **exotic**
  a descriptive term for animals or plants introduced into an area or country from outside or abroad, not naturally occurring here.

- **exotoxins**
  1. soluble toxins disseminated out of living bacteria into the environment.
  2. toxins secreted by living micro-organisms.

- **expanded**
  as applied to feed having been increased in volume as the result of a sudden reduction in surrounding pressure.

- **expectation**
  the expected value of a function of variate value is its mean value in repeated sampling.

- **expectoration**
  the act of spitting out materials, as in evaluating dairy products when it is unwise to swallow samples because of accumulative effects on the ability to evaluate product quality.

- **expeller process**
  a process for taking out the oil from oilseeds in which the seeds are forced along a tapering tube, squeezing out the oil. The residues containing usually less than 5% oil, maybe used as animal feeds.

- **experiment**
  from the Latin experiment urn meaning proof from experience. It is a procedure used to discover or to demonstrate a fact or general truth.

- **export subsidies**
  government subsidies on exports to make them more competitive in international trade.

- **exsanguinations**
  the extraction of blood, as by the blood sucking insects.

- **extender**
  a substance added to a sample of semen to extend its viable life-span.

- **extirpation**
  the complete taking away or annihilation of a part.

- **extract**
  a preparation obtained by soaking a crude drug in a liquid solvent. The dissolved active ingredient is separated from the inert mailer and concentrated or evaporated until almost or completely dry.

- **extrinsic factor**
  a factor coming from or originating from outside an organism.

- **extroversion**
  the method of exposure of tissues inside the streak canal of cows' teats usually caused by prolonged milking a ad excessive milking vacuum.

- **extruded**
  as applied to feed having been forced through a die under pressure.

- **exudate**
  an abnormal seeping of fluid through the walls of vessels into adjacent tissue or space.

- **exudation**
  the escape of fluid, cells and cellular debris from blood vessels and their deposition in or on the tissues.

- **exudative diathesis**
  symptom of vitamin E deficiency in poultry characterised by an gathering of fluid in subcutaneous fatty tissue.

- **eye level parlour**
  a milking machine where the recorder jars are hung at eye level.

- **face**
  the region of bare skin around and below the eyes.

- **faces**
  excrement discharged from the intestines.

- **facet**
  a small articular facade.

- **factor**
  a situation affecting the result of an observation or experiment. It is used especially for an act of related treatments in an experiment.

- **factorial experiment**
  an experiment in which all the treatments or agents under investigation are varied simultaneously and combined in such a way that any

derived effect of one or group of item may be isolated and evaluated separately.

- **facultative**
having the power to live under diverse conditions, as with bacterial growth in both high and low concentrations of oxygen.

- **fall**
1. the dropping of an animal from us mother's uterus to the ground during birth.
2. also the number of offspring born in a herd of flock, etc. in a season.

- **fall wool**
wool sheared in the fall in the parts of the country where sheep are sheared twice a year, ii is usually of lour to sis months growth and generally inferior to spring wool.

- **fallen wool**
a type of wool oddment which is either clipped from a dead sheep, or is badly shorn and is in the form of torn or broken pieces as distinct from an intact fleece.

- **far side**
the right side of a horse, in horsemanship.

- **farm**
1. an area of land under single ownership and used for agriculture either to raise the crop or pasturage.
2. any place that raises few numbers of livestock or a few hectares of crops

- **farm enterprise**
any crop type or livestock, manufacturing process etc. which comprises a parity of the farm business.

- **farm processing**
any process or treatment used to get ready farm products for use or preservation. It includes the procedure that maintains or improves the quality of farm products or changes their forms like grain grinding, chaff cutting, grain drying, cane crushing , seed processing, milk processing etc.

- **farm retail price spread**
difference between the price paid by consumers for one

unit of a food product and the farm value of an equivalent amount of the product.

- **farm yard**
  a small area which is adjacent to a cattle stall and in which animals are kept when not in pasture, at work or in the stall.

- **farmer's lung**
  a distressing respiratory condition caused by the inhalation of dual particles, e.g. from mouldy feed or litter, containing fungal spores, which results in an allergic reaction in the air sacs of the lung and causes breathing difficulties cows and horses can also suffer.

- **farmstead**
  a farm house and the farm buildings associated with it.

- **farrier**
  a person who makes and fits horse shoes.

- **farrow**
  1. an expressive term for a cow or heifer not in calf or barren
  2. to give birth to a litter of pigs, also the litter itself.

- **farrowing**
  parturition or delivery of piglets at the end of pregnancy.

- **farrowing crate**
  a steel enclosure used in a pig pen to surround a sow and prevent it from lying on and killing the piglets; whilst allowing them access to her teats. They are usually in the form of a long thin rectangle, but are sometimes circular (the Ruakura pen). A *farrowing rail* serves the same purpose and is usually fitted a short distance from the pen walls and floor.

- **fat**
  1. a group of nutrients, each usually made up of glycerol and three fatty acids. The most powerful energy source in rations. Contains the elements carbon, hydrogen and Oxygen, Hydrogen present in much higher proportions than in Carbohydrate.
  2. the product formed when a fatty acid reacts with glycerol. The glyceryl ester of a fatty acid. Stearin, palmitin, aud olein are examples.
  3. a term applied to animals reared for their meat (e.g., fat cattle, fat sheep) which

have been fattened up and are in a ready condition for sale to a butcher at a market.
4. a dry measure of nine bushels.

- **fat class**
a visual appraisal of the degree of external fat development on a beef or sheep carcase, being one of the characteristics assessed in carcase classification. There are five main classes ranging from 1 (very lean) to 5 (very fat). For sheep, class 3 maybe sub-divided into 3L and 3H; whilst for beef elates 4 to 5 may be sub-divided into 4L and 4H and 5L and 5H.

- **farrowing rate**
the number of sows actually farrowing as a percentage of sows served initially.

- **farrowing interval**
the average time for a sow or a group of sow between one farrowing and the next.

- **fascia**
this is a thin sheet or band of fibrous (connective) tissue which supports, covers, or binds together a muscle, part or organ.

- **fastidious**
in microbiological context, it means an unusual requirement for preformed nutrients or specific environmental conditions.

- **fasting**
abstain from food. Certain types of experiments involve withholding food from the test animal for varying periods of time.

- **fat corrected milk**
adjustment of milk with different fat percentages to equivalent amounts on an energy basis.

- **fat differential**
the amount added to or subtracted from the combine price for each 0.1 per cent that the milk fat is above or below 3.5 per cent.

- **fat globules**
the naturally occurring spheres of butter fat in milk.

- **fat soluble**
those substances which are soluble in fats and fat solvents but generally not soluble in water.

- **fat soluble vitamins**
vitamin A, D, E and K.

- **fatling**
  a young animal fattened for slaughter.
- **fats**
  storage materials found in living organisms, chiefly in solid form (also called adipose tissue) in animals and in liquid form (oils) in plants, comprising mixtures of glycerides (condensation product of glycerol and fatty acids). most animals deposit fax in the body as an energy store. some plants, mostly tropical, produce seeds rich in oil which is extracted mainly for cooking, the residue being used for cattle cake.
- **fatstock**
  these are livestock fattened for sale in a market.
- **fattening**
  the deposition of unused energy in the form of fat within the body tissues.
- **fatty acids**
  the organic acids, thirteen of which are found in natural fats. They are either,
  1. saturated (i.e., molecules to which no further atoms can be added), found mainly in solid form in animal fats, or
  2. unsaturated, found mainly in liquid form (oils) in plants. Common found fatty acids in natural fats are palmitic, oleic and stearic (Steapsin).
- **fauna**
  the animal life present. Usually used to refer the overall protozoan population present in an ecology.
- **faverolle**
  a weighty table breed of poultry, with a single comb and a little feathered white legs bearing five toes. It produces white eggs. Variously coloured including white, salmon, buff and blue.
- **favour**
  to use or to protect carefully; as an animal favours a lame leg.
- **FCM (4 per cent fat corrected milk)**
  the process of evaluating milk-production records of different animals and breeds on a common basis energy-wise. The formula it uses is 0.4 x milk production + (15 x pounds fat produced).
- **FCM**
  It is 0.4 x milk production +

(15 x pounds fat produced).

- **feather**
  1. long hairs found on the sides and on the back of the legs of some breeds of horse e.g., the shire.
  2. one of the dermal outgrowths forming the plumage of a bird.

- **feather legged**
  legs, which have feathers at the sides of the shanks and feet.

- **feather meal**
  the product which is derived from the treatment under pressure of clean, un-decomposed feathers from slaughtered poultry free of additives and/or accelerators. The condition is that not less than 80% of its crude protein content must consist of 'digestible protein'.

- **feather tracts**
  the fussy regions on a fowl's body on which the feathers grow.

- **febrile**
  pertaining to a fever, or rise in body temperature.

- **fecal energy (FE)**
  the food energy which is lost through the faeces.

- **faeces**
  the excreta which is discharged from the digestive tract through the anus.

- **fecundation**
  being fertilised or pregnant.

- **fecundity**
  the level of reproductive effectiveness attained in terms of farrowing index.

- **fee**
  livestock, particularly cattle (a backdated term for).

- **feed**
  stuff which are eaten by an animal as a part of its daily ration.

- **feed block**
  a block of foodstuffs left on pastures, especially in hill areas, for sheep to lick at will as a support for maintaining conditions. Usually, they contain a carbohydrate source (e.g., cereals, glucose:, or molasses), protein and urea (sometimes excluded), minerals, vitamins and trace elements.

- **feed conversion (feed efficiency)**
  the units of feed consumed per unit of weight increase. Also, the production (meat, millet and eggs) per unit of

feed consumed.

- **feed formula**
the method in which two or more food ingredients are mixed and processed according to requirements proportionately,.

- **feed grade**
a grade of food suitable for animal but not for human consumption.

- **feed hopper**
a through pot or pan for keeping food for poultry birds. Usually it remains protected by running wire cross-wise to prevent birds from scratching out food.

- **feed ingredient**
components or materials which constitute a commercial feed.

- **feed off**
a practice of allowing a crop to be eaten by animals when the crop is still in the ground.

- **feed ring**
a raised circular container chiefly used for hay and straw from which livestock feed around its circumference.

- **feed supplement**
a feed which is used with another to improve the nutrient balance at the total and intended to be:
1. feed dilutent as a supplement to other rations;
2. offered free choice; 01
3. further diluted and mixed to produce a complete feed.

- **feedback inhibition**
the inhibition of the first enzyme in a metabolic pathway by the end product of that pathway.

- **feeder**
a young animal that is not highly fatty but show, evidence of ability to add weight economically.

- **feeding index**
the percentage relationship of amount of food reported fed to amount needed (on an estimated *net energy* basis). The required amount needed includes proper allowances for body maintenance, milk production at the average herd level, growth for cows under four years of age and pregnancy requirements in late lactation and dry period.

▪ **feeding standards**
these are the tables showing the amount of food and nutrients which should be provided in ration.

▪ **feeding stuffs**
varieties of foods available for farm animals. These can be classified into four main groups, viz.,
1. concentrates.
2. dry roughage or coarse fodder, e.g., hay derived from grasses, cereals and legumes, or straw derived from cereals and the haulms of peas and beans and
3. succulent foods, mainly roots and tubers, e.g., turnips, potatoes etc.,
4. green forages,

▪ **feedlot**
an region of land in which animals are put up at very high density. This area of land does not contribute at all to the production of animal food, all of which has, therefore, to be brought to the animals from outside the feedlot.

▪ **feedstuff**
any material included in diet because of nutritional properties.

▪ **fell**
1. the hide of skin of an animal, mainly when covered with hair.
2. a fleece.

▪ **fellmonger**
a person who prepares animal skins for tanning.

▪ **fellmongering**
a chemical process to take away wool from the hide by bacterial action.

▪ **felting**
the property possessed by wool and some other animal fibres of closely mingling and interlocking to farm a compact mass.

▪ **felting property**
the property in wool which causes the fibres to lock together when subjected to the action of heat, moisture, alkali and friction.

▪ **femininity**
a specific physical appearance resulting from well-developed secondary female sex characteristics.

▪ **femoral**
related to the femur or to the thigh.

- **femur**
  the thigh bone.

- **fencing**
  a tool that fixes the boundary of the farm and helps to stop the encroachment of the land and protection from wild animals and cattle.

- **feral**
  the domesticated animals that have reverted to the wild state.

- **ferment**
  an enzyme which is capable of producing fermentation of the other substances with which it comes into contact, e.g., yeast.

- **fermentation**
  the chemical changes which takes place due to enzymes produced by various micro-organisms.

- **fermented milk**
  cultured butter milk. Milk seeded with cultures of lactic acid bacteria to bring about souring and to cause changes in texture, aroma and taste. A refreshing beverage.

- **ferredoxin**
  a group of red-brown protein which contains iron in association with sulphur. They have negative redox potentials. They are strong reducing agents. They are common components of electron transport chains in most organisms.

- **ferret**
  this is an albino variety of the polecate which is half-tamed and used to catch or unearth rabbits and rats.

- **fertile**
  able to produce offspring.

- **fertility**
  the reproductive effectiveness attained in terms of productiveness and prolificacy.

- **fertilisation**
  the union of the male and female gametes to form a new individual. In this process two haploid cells are combined to restore the diploid number of chromosomes in the new individual.

- **fet**
  the term refers to the progeny of a ram; normally applied to its offspring in one particular season.

- **fetlock**
  the clump of hair which

grows above a horse's hoof or the part of the leg.

- **foetus**
period in the development of an individual between completion of organ formation and birth.

- **fever therapy**
the treatment of disease by causing the development of lever.

- **fibre**
a constituent of animal feeding stuffs, mainly contains cellulose and lignin which aids digestion by stimulating muscular activity in the digestive tract. It also facilitates the digestion of concentrates by opening them up to the action of the digestive juices. All farm animals require a certain amount of fibre except when very young and on a liquid diet. Ruminants require the most fibre and derive energy from it by bacterial digestion in the rumen. cows require adequate fibre to produce milk with a high percentage of butterfat. sheep and horses require less fibre than cattle bill more than pigs. excess fibre in the diet can prevent an animal, eating sufficient food to derive adequate essential nutrients hay and straw are characterised by a high fire content.

- **fibril**
a single fine thread structure (in the dermis).

- **fibrin**
a substance upon which depends the formation of blood clots.

- **fibrinogen**
a soluble protein found in the blood and body fluids of animals which is essential to the coagulation of blood.

- **fibroblasts**
connective tissue cell's whose function is one of repair.

- **fibroma**
a hard papule, nodule or tumour which histologically shows fibroblasts and their product (collagen bundles) in a dense condition.

- **fibrosis**
a condition marked by the presence of interstitial fibrous tissue, particularly in the mammary gland resulting from mastitis.

■ **fibrous**

constituted with fibres or strands.

■ **fibrous carbohydrates**

the cellulose and hemicelluloses feed components, which are not readily digested by animals.

■ **fibula**

the name of the smaller of the two bones connecting the knee and ankle.

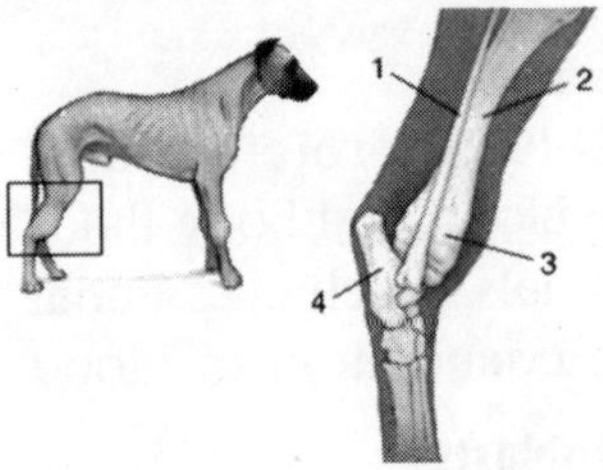


■ **fifth toe**

a few breeds have an extra toe at the rear apart fro the four toe which is found in most of the animals, for example, Dorkings and Faverolles. This extra toe is called the fifth toe.

■ **fill**

a term designating the fullness of the digestive tract of animal. also used to identify a phase of packaging milk and/or milk products in which containers are filled with product.

■ **filled milk**

milk from which milk fat has been removed and replaced with other fats or oils.

■ **filled products**

milk products in which milk fat has been replaced with another fat or oil, usually vegetable origin.

■ **fillers**

the inert ingredients, used to increase the weight of feed, drug or fertiliser preparations or to modify their properties.

■ **filling**

the weft, that is, the yarn that runs the short way of a woven fabric; usually made of thicker, softer yarns than the warp; hence weaker and shorter wool can he used for it than for warp.

■ **filly**

a young female horse between 1 and 4 years of age.

■ **filly foal**

a female horse up to 1 year of age.

■ **film badge**

an appropriately packaged photographic film for detecting radiation exposure of personnel; usually dental-size x-ray film, worn on the person and frequently combined with an identification badge.

- **filterable virus**

a dame commonly applied to a pathogenic agent capable of passing fillers that retain bacteria.

- **find**

lay term meaning to give birth of young, e.g. , a cow finds a calf.

- **fine wool**

1. a common term applied lo wool from the various sub-breeds of merino sheep. 2. the finest of i he united stales official wool grades as established by the u.s. bureau of agricultural economics.

- **finely drawn**

applied to tall birds with fine bones, spare body, light feathers and narrow sickles, light and sprightly.

- **fineness**

the smallness of diameter of the individual wool fibre.

- **finish**

refers to fatness; highly finished means very fat.

- **first calı**

the first calf born to bovine females.

- **first calf heifer**

a term used in same region for a heifer after having borne its first calf but before it has borne a second calf.

- **first cross**

the offspring from the mating of two different pure breeds, possessing hybrid vigour when egg or table production is the aim. toe breed of the male is given first, for example. Rhode island red x light susses.

- **first meiotic division**

the first of two divisions occurring in reduction cell division and resulting in the production of two cells, each of which is haploid, the chromosomes occurring as paired chromatids joined at the centromere.

- **first service conception rate**

the percentage of cows in a herd in which pregnancy is established after the first service.

- **fish meal**

an animal food stuff consisting ot dried and ground waste or unsold fish and fish filleting residues, by jaw containing not more than 6% oil and 4% salt, white fish meal is derived from non-oily 'while fish, (e.g. cod, haddock, etc.) and usually contains about 66% protein. other fish meals are derived from oily fish (e.g.. herring, pilchards, etc.) alter most of the oil has been extracted and are usually richer in protein than white fish meal. Fish meal is mainly fed to pigs and poultry, but is also included in rations for dairy cows, calves and other farm animals.

- **fish oil**

oil extracted from un-decomposed dried fish.

- **fish scrap**

consists mostly of un-edible piece like bones and some intestines of fish,

- **fish soluble matters**

products obtained by concentrating the 'slick-water' or clarified liquor, which consists of dissolved, dispersed fish protein particles along with minerals, B-complex vitamins and some unknown growth factors. The material is obtained as a by-product during fish meal production,

- **fisher, Sir Ronald**

a British biometrician (1890-1962) who developed much of the underlying theory of modern statistical genetics and biological statistics.

- **fission**

the split of a mature nucleus into two approximately equal parts, accompanied by the release of a fairly large quantity of energy and in general one or more neutrons.

- **fissure**

a splitting or discontir uity of a surface, specially one that persists.

- **fleece**

1. the wool of one sheep considered as a unit;
2. the pan of a skirted fleece that is left after the skins have been removed. the wool in the fleece is supposed to cling together in one piece.

- **flexion**
  the bending of a jointed structure, e.g., fingers.

- **flights**
  the principal leathers of the wing used in flying, but folded out of sight when not in use.

- **flitch**
  a side of bacon, sailed and cured.

- **float**
  a low cart used to carry livestock to produce,

- **flock**
  1. a jocks or tuft wool (cotton or hair), especially when waste is used to stuff upholstery, mattresses, etc.
  2. a company or compilation of animals. particularly applied to sheep, goats and birds.

- **flock book**
  a record for a particular breed of sheep of registered sheep conforming to the breed type and their pedigree relationships.

- **flock mating**
  mating of a number of females and males together in one flock.

- **flock ram**
  a pure-bred ram of not as much high standard as a stud ram used for mating with flock ewes.

- **flock master**
  the person in charge or owner of a hack of sheep.

- **flocks**
  incredibly short fibres produced in the milling and finishing of cloth. They are also produced by chopping up rags and tailors' clippings. They are used to some extent for adding weight to fabrics with a felted face and are also used in paper manufacture.

- **flora**
  the bacteria in or on an animal usually referred to as bacterial flora, as in the digestive tract.

- **fluff**
  1. the soft plumage on the thighs and abdomen of an adult fowl; also the soft fluffy plumage at the base of a feather.
  2. the original dawn of day-old chicks.

- **fluid milk**
  milk usually marketed as

fresh liquid milks and creams. It is the most perishable form of milk and commands the highest price, per unit volume. Also called market milk.

- **fluid milk products**
  milk, skim milk, filled milk, concentrated milk , flavoured milk , cream products , low fat milk drinks and fortified skim milk.

- **flukes**
  parasitic flat worms found in the lung, liver, guts and blood vessels of animals. liver flukes are especially common in sheep and cattle.

- **flukey area**
  a poorly drained grazing land where mud-snails are found, which are the alternative host in the transmission of liver fluke disease.

- **fluorosis**
  also called *fluorine poisoning*, a chronic and insidious disease of cattle resulting from long-lasting intake of fluorine. It is characterised by lameness, stunted growth , thickening of jaws, mottled and irregular teeth and bones of limbs.

- **flush**
  1. to give an animal a better off diet prior to mating for improving its condition.
  2. a sudden, generally rapid growth of foliage, usually applied to grass.

- **flush season**
  time during the year usually spring and early summer of maximum milk production.

- **flushing**
  the method of high plane energy feeding for a short time to supplement ovulation rate.

- **flux**
  the number of neutrons passing through one square centimetre of a given target in one. It is a measure of the intensity of neutron radiation.

- **fly strike**
  the state produced by the growth of blowfly maggots on the living sheep (*cutaneous myiasis*).

- **fly wool**
  pieces of loose wool that fall from a fleece during shearing .

- **fomites**
  all articles that have been in

actual contact with a sick animal, so as to retain some of the infective material and be capable of spreading the disease.

- **food and fodder register**
  it records only the consumption of cattle feed (purchased) from outside and produced on the farm and fodder by the livestock of the farm and day-today consumption is written in this register.

- **food chain**
  a series of plant or animal species in a community, each of which is related to the next as a source of food. For instance, plant is consumed by animals and the waste products of animals are again consumed by plants from the soil, thus forms a food chain.

- **food conversion ratio**
  the number of kilograms of food consumed by an animal required to produce a live weight gain of 1 kg, a measure of the efficiency of the animal in converting food into flesh, a small ratio indicating high efficiency.

- **food poisoning**
  a general term applied to all stomach or intestinal disturbances due to food contamination with certain micro-organism or their toxins.

- **food unit**
  1 kg. of average barley, equivalent to 0.71 kg. starch equivalent.

- **foot and mouth disease**
  a extremely infectious viral disease of cattle, sheep, goats and pigs, characterised by the development of blisters ill the mouth. causing considerable salivation and on the feet resulting in lameness. Death is exceptional, but animals stop gaining weight and milk production in dairy cows drops down. Epidemics occur from time to time and the disease has been endemic in many parts of the world, the disease can be spread by infected bones, offals, birds, etc. and by the wind. Diseased animals and their contacts are generally slaughtered.

- **foot candle**
  the intensity of illumination on a surface when a light

source, emitting 1 candela of luminous energy, is placed 1 fort away from it.

■ **foot rot**
a chronic wound-infection which has been characterised by extreme lameness and frequently by distortion of the hoof. It occurs due to irritation by mud which dries and cracks the skin, by sand and by gravel between the claws, animals badly affected may walk on 3 legs.

■ **forage**
vegetable feed for domestic animals. Generally, the term refers only to such material as pasturage, hay, silage and green feed; as in contrast to less digestible plant known roughage.

■ **forage box**
a huge movable container, usually with wheels, similar to a dump box, used to transport scavenge from a Silo to a manger. The forage is discharged in a steady even flow, usually by a transverse moving floor mechanism or elevator bell, driven by p.t.o. from a tractor. Also from time to time used to transport forage from the field to mangers for flocks under zero-grazing.

■ **forage crop**
plants grown primarily for livestock feed, to be either harvested for hay, silage, or green feed or harvested by grazing animals.

■ **forage harvester**
a machine which chops, cuts and loads (by chute) green crops (e.g., grass, lucerne, etc.) into an adjacent or towed trailer, for subsequent silage making. Sometimes the crop is precut and allowed to wilt in windrows before being collected by a pick-up reel substituted for the cutter mechanism. There are three main types, viz. Double chop harvester, flail forage harvester and metered chop harvesters. They are mainly trailed or semi-mounted but self-propelled machines are available.

■ **foramen**
an opening or hole. The word is applied particularly to holes in bones through which pass blood-vessels or nerves.

- **forceps**
  a pliers-like tool used in medicine and certain research studies for grasping, compressing and pulling.

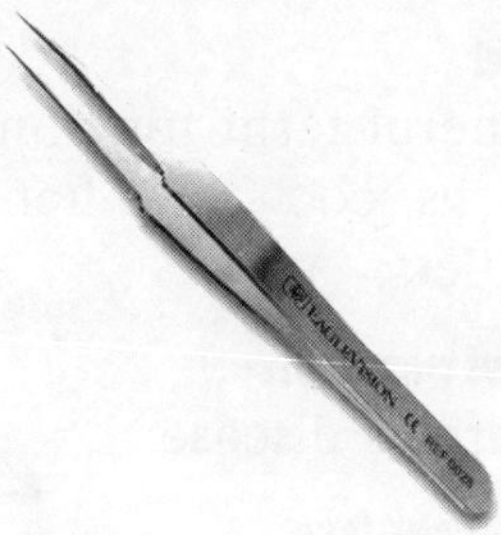

- **foremilk**
  the first few squirts of milk drawn from a cow's udder by hand into a strip cup before starting milking. this milk has a high bacterial content
- **forequarters**
  the front two quarters of an animal.
- **forage blower**
  an apparatus used to fill a Silo, mainly of the tower type, with finely chopped material for making silage, by blowing it up a duct, often to a height of 15m (50 ft.) into the Silo. Some kinds of forage blower also chop the crop, these being called *cutter blowers*.
- **formic acid**
  a colourless pungent compound.
- **formula feed**
  two or more ingredients mixed and processed according to requirement proportionately.
- **fors**
  hairs in fleece, mixed with the wool. the taking away of such hairs from wool is called *forsing*.
- **fortified cod-liver oil**
  cod-liver oil mixed with a small quantity of other fish oil, fish oil concentrates or vitamin D.
- **fortify**
  to add one or more nutrients to a food so that it contains more of the nutrients than was originally present before processing. Milk often is fortified with vitamin D.
- **forward**
  a descriptive term applied to both livestock and crops that are more advanced in their development at a particular time than usual.
- **fossa**
  a depression — a concavity.

- **foster mother**
  an ewe rearing a lamb other than her own.

- **fuelein the foot**
  a bacterial infection (generally actinomyces necrophorus of the hooves, particularly of the hind feet, of cattle, normally around wounds or cuts. swelling, heat, pus production and lameness occur. muddy fields and gateways are particular sources of infection.

- **foul sheath**
  a venereal disease in rams.

- **fouled**
  those pasture spots or areas made improper to the grazing animal with the presence of urine or dung.

- **founder**
  inflammation of the fleshy laminae within the hoof, from concussion, overfeeding and many factors causing an over-supply of blood to this area of the hoof. It causes great pain to the affected individual. also called *laminitis*.

- **four tooth sheep**
  sheep of about 18-21 months of age.

- **four way cross**
  offspring from the mating of two single crosses.

- **fovea**
  a pita noticeably concave depression.

- **fowl**
  general term for domestic cocks, cockerels, hens and pullets.

- **fowl paralysis**
  Marek's disease.

- **fowl cholera**
  a contagious bacterial disease (*pasteurella apisepticus*) of poultry, characterised by sudden high fever and profuse green diarrhoea.

- **fowl pest**
  an acute, highly fatal, infectious disease of chicken, turkeys and peasants and ducks and other water birds. It is caused by a filterable virus having a particle size of 60-80m diameter.

- **fowl pox**
  a viral disease of fowls causing warts on the comb, wattles and other parts of the head.

- **fowl typhoid**
  an infectious bacterial disease (salmonella gallinarurn) of

poultry, mainly affecting pullets, causing drowsiness and loss of appetite.

- **fox**

a carnivorous canine predator (*Vulpes*) known for its habit of taking young lambs and poultry and is a vector of rabies.

- **fox trot**

an easy, shun, broken gait of a horse, between a walk and a. trot.

- **fractionation**

the laboratory separation of natural materials into their component parts.

- **free**

a term applied to wool that is free from vegetable fault.

- **free choice**

a feeding system by which animals arc given unlimited access to the separate components or groups of components constituting the diet so that they can choose the portions of each they prefer.

- **free grazing**

the must common method of grazing stuck with few, if any, controls. stocking rate is usually adjusted to the supply of herbage but over- and under-grazing often occurs. Fencing cost are kept to a minimum and intensive stock super-vision is avoided (grazing systems).

- **free radical**

an atom or group of atoms having at least one unpaired electron- they participate in oxidative reactions in some dairy foods.

- **free range**

a arrangement of keeping poultry in which the birds are provided with small houses and are allowed to run free over a field or large enclosed area. Each house is usually designed to accommodate 50-150 birds and is raised slightly above the ground. Most poultry are now-a-days kept under more intensive systems such as battery cages.

Pigs may also be kept under a free range system, with breeding sown provided with cheap mobile housing or arks for protection and having the run of an area of grass. the majority of pigs are also kept under more intensive systems.

- **free stalls**
compartments and similar housing arrangements in which dairy animals are free to enter and leave, as opposed being confined in stanchions.

- **freemartin**
an infertile female calf born co-twin to a bull; the abnormal genital tract development present in the female is the result of circulatory connections between the two sexes of calves in utero; approximately 90% of the female members of female/ male twin sets are freemartins.

- **french combing**
an middle length between strictly combing and clothing wool. It is so named because French combing machinery can comb shorter wool than the machinery that was for a long time the standard in England.

- **frenching**
a disease characterised by loss of colour in leaves between veins.

- **frequency distribution**
method of depicting the proportion of animals whose phenotypes occur at various points along a horizontal axis; the height of the line drawn over the scale representing the frequency of occurrence at each point on the scale.

- **fresh**
1. the green or wet form of a feed material
2. a cow that has recently given birth to a calf.
3. dairy products having original qualities unimpaired to calving (parturition):
4. to give birth to a calf and concurrently initiate lactation.

- **freshen**
to give birth to a calf and simultaneously to begin a period of lactation.

- **fribs**
short tuffs of wool frequently called *second cuts* because they are produced by the staple being cut twice in the process of shearing.

- **friedman test**
a teat for pregnancy in which a small amount of urine of the tested animal is injected into the blood stream of a virgin rabbit.

Pregnancy is indicated by certain changes in the ovaries of the rabbit.

- **friesian**

a breed of black and white cattle originating in Holland and exported across the world, so that tome variations now exist. The British Friesian is intermediate between the well -muscled, short-legged, dual-purpose Dutch Friesian and the Holstein Friesian (Holstein) of Canada which produces a high yield of milk and Butterfat. It is characterised by longer legs than the Dutch Friesian, with a deep body, long and wide hindquarters and a urge udder. the average annual milk yield in 1981/82 (milk yields) was 5885 kg. (5703 litres) with a 3.83% butterfat content. red and white types are also known,

- **frizzled**

feather which is naturally curled backwards over itself, for example, the Frizzle fowl.

- **full feed**

the term commonly applied to faltering cattle being provided as much as they will consume safely without going off feed.

- **full mouthed**

a descriptive term for livestock in which a complete set of permanent teeth have grown, e.g., after 3 years in sheep.

- **full sibs**

animals having the same sire and dam, such as full brothers, full sisters.
A full brother-sister pair.

- **fulling**

the process of recoiling and felling woolen cloth so that it becomes thicker and denser and has a face produced upon it which covers up the weave pattern.

- **fumigant**

a liquid or solid substance that forms vapours that destroy pathogens, insects and rodents.

- **fulminant**

happening abruptly and with immense severity.

- **function disorders**

those which cannot be attributed to lesions of organs or tissues; as compared with 'organic' disorders where lesions are evident.

- **funds**

the base or innermost part

of a hollow organ distant from its opening.

- **fungi**

plants that contain no chlorophyll, flowers, or leaves. They get their nourishment from dead or living organic mailer.

- **fungicide**

an agent that destroys fungi.

- **fungistatic**

an agent or process that arrests the development of fungi.

- **furnished**

a cock when he has a full tail- hackles and comb, etc, is considered furnished.

- **furrowing**

the process of cytokinesis in animal cells.

- **furuncle**

a sore, nodutar abscess in the skin caused by bacteria; a boil.

- **fusion**

the formation of a heavier nucleus from two lighter ones (such as hydrogen isotopes), with the attendant release of energy (as in a hydrogen bomb).

- **futures market**

a market at which contracts for future delivery of a commodity are brought and sold.

- **gadfly**

a name given to a number of blood sucking flies but mainly the *Tabanidae,* which attack horses, cattle and other animals. The European gadfly *Tabanus bovinus* is about 25 mm. long coloured black and yellowish brown; other common species are about half of this length. The female sucks the blood of horses and cattle, attacking them in vulnerable parts and causing horses. some species are carriers of Trypanosomes and predatory, live mainly in wet situations, in mud or under shallow water. They move to a drier place to pupate.

- **gadsman**

a Scottish term for a person who drives horse-drawn ploughs.

- **gait**
  the style of walking. Defects in gait are because of damage in the spinal cord or brain.

- **galactophore**
  a milk duct.

- **galactophorus**
  carrying or producing milk; conveying milk.

- **galactopuietic**
  that which stimulates or increases the secretion of milk.

- **galacloirhoea**
  an excessive secretion of milky fluids from the breast, especially when not associated with pregnancy.

- **galactose**
  carbohydrate which in amalgamation with glucose forms lactose.

- **gall bladder**
  the little pouch-like sac in which bile produced by the liver is stored until it is required during the process of digestion.

- **gallstones**
  are spongy, brownish foreign bodies consisting of cholesterol bile pigment and lime gallstones often causes jaundice.

- **Galton Sir Francis**
  British statistician (1822-1911) who is considered to be the father of the science of eugenics, i.e., the application of statistical theory for genetic improvement.

- **galvanise**
  to coat with zinc.

- **gambrel**
  a bracket-shaped piece of iron on which a carcass of a sheep is hung by its hind legs during the process of dressing or when hung in a butcher's shop.

- **gamma globulin**
  protein fraction of blood serum especially high in antibodies.

- **gamma rays**
  high-energy short-wavelength electromagnetic radiation. Gamma rays are very penetrating and are best stopped or shielded against by dense materials, such as lead. Gamma rays are similar to x-rays but are usually more energetic (their penetrating power is greater) and are of nuclear origin.

- **gander**
  a mature male goose.
- **ganglion**
  an aggregations of nerve cells forming a mass of grey, nervous substance and serving as a centre of nervous influence
- **nang mowing**
  mowing by 3 or 5 linked cylinder-mowers at the same time to give a very close cut.
- **gapes**
  a disease of young chickens and turkeys particularly, although all the domesticated and many wild birds may also be affected, which is due to the presence of small worms in the bronchial tubes.
- **gargot**
  inflamed state of tissue just in front of udder at time of parturition. Erroneously applied to thick, stringy milk found in mastitis affected cows.
- **garneting**
  the method of improving wool fibres from rags, etc, by tearing them up between oppositely moving toothed rollers.
- **gas gangrene**
  an sensitive bacterial disease due to the immunization of wounds with organisms belonging to the 'gas gangrene' group,
- **gaseous products Of digestion**
  include the combustible gases produced in the digestive tract incident to the fermentation of the ration. Methane constitutes the major proportion of the combustible gases produced by the ruminant, however, non-ruminants also produce methane. Trace amounts of hydrogen, carbon-monoxide, acetone, ethane and hydrogen sulphide are also produced.
- **gastrectasis**
  dilatation of the stomach.
- **gastrectomy**
  an process for the removal of the whole or part of the stomach.
- **gastric**
  anything connected with the stomach, e.g., gastric ulcer, gastric juice.
- **gastric juice**
  a clear liquid secreted by the wall of the stomach, it con-

tains hydrochloric acid and the enzymes rennin, pepsin and gastric lipase.

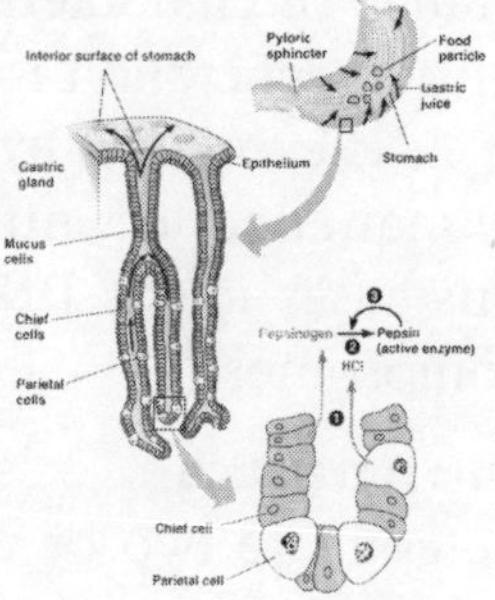


■ **gastritis**
an inflammation of the stomach, especially of the lining, or mucous membrane.

■ **gastroenteritis**
inflammation of the mucosa of the stomach and intestines.

■ **gastrula**
an early stage in embryonic growth; it follows the blastula stage.

■ **gavage**
the force feeding of poultry, generally with the aid of a stomach tube.

■ **gay**
the high or 'squirrel' tail, also the heavy white splashes on Anconas, Houdans, Exchequer Leghorns, Hamburghs and Spangled Game.

■ **geiger counter**
tool that counts pulses produced by radioactivity, consisting of a counting tube with a central wire anode, generally filled with a mixture of argon and organic vapour.

■ **geld**
to castrate (castration), especially a horse.

■ **gelding**
a castrated male horse. To geld is to make sterile or to take away the testicles. Horses are generally castrated at about two years of age.

■ **gene**
the primary unit of inheritance. Physically, each gene is apparently a nucleic acid with a unique structure. It influences certain traits. Sometimes called a *trait determiner.*

■ **gene ecology**
a branch of biology that deals with the genetics of population of animals and plants in relation to the ecological niches they occupy.

- **gene frequency**
proportion (percentage) of a certain gene in its allelic series.

- **generation**
all those individuals of an organism which has a common parent or ancestor by an equal number of reproductive cycles.

- **generation interval**
the time from the birth of one generation to the birth of the next generation. in humans, it is calculated by subtracting the average age of the children from the average age of their parents.

- **generation length (interval)**
standard length of time from birth of one generation to birth of the succeeding generation.

- **generation time**
the time necessary for growing cells to double their numbers or mass.

- **genetic**
pertaining to heredity.

- **genetic base**
the average genetic merit of a population (usually a breed) at a specified time used as a reference pain to estimate of genetic transmitting ability.

- **genetic code**
the code controlling the inheritance of characteristics from one generation to the next. It is expressed by linear sequence of nitrogenous bases in the DNA of chromosomes.

- **genetic correlation**
a state where two or more quantitative traits are determined by many of the same genes.

- **genetic determiner**
the gene.

- **genetic engineering**
the replacement of a gene (usually defective) by a gene from an outside source. introduction of a gene from one bacteria into another.

- **genetic equilibrium**
the state of a population in which succeeding generations consist of the same genotype with the same frequencies, in respect of particular genes, or arrangement of genes.

- **genetic information**
the information contained in a sequence of nucleotide

bases in a DNA or RNA molecule.

- **genetic map**
  arrangement of mutable sites on a chromosome which has been deduced from genetic recombination experiments.
- **genetic sterility**
  a kind of male sterility conditioned by nuclear genes in contrast in cytoplasmic sterility. It may be transmitted by either the male or female parent.
- **genetic transmitting ability**
  the standard genetic merit for some specified trait that is passed on by a parent to its offspring one-half breeding value.
- **genetic value**
  the sum value of the genes an individual for performance, type, or carcass quality and quantity.
- **genetic variation**
  variation due to differences in gametes.
- **genetics**
  the science of heredity, including the study of its chemical foundation, its development, expression and its bearings and variation, selection, adaptation, evolution, breeding and the activities of man.
- **genie balance**
  the hypothesis that the characters of an organism are each determined by the interaction of a large, but unknown number of genes, some affecting developing in one direction and some in another, so that ultimate result is a balance struck between the total effects.
- **genital**
  female sex organ used for reproduction.
- **genital papilla**
  a visible projection inside the vent of the day-old male chick which distinguishes the male from the female chick which lacks it.
- **genitourinary**
  refers to the organs of reproduction and urine excretion.
- **genome**
  the total genetic composition of an individual or population which is inherited with the chromosomes. a haploid set of chromosomes with their genes.

- **genophore**
  the strand of DNA that bears the genes and occurs within the nucleoid of procaryotic organisms or in virus particles.

- **genotype**
  the genetic constitution of an organism, in terms of the particular combination of characters each of its cells inherits from the parental cells. not necessarily the characters manifested by the organism.

- **genotypic environment**
  the collective of all the genes, considered as acting on one or more of them.

- **genotypic selection**
  the choice of animals in breeding on account of their genotype, shown by their performance following close study of their breeding behaviour and the results of progeny testing.

- **gentian violet**
  a stain used in microscopically work and a valuable antiseptic, of use against fungus and bacterial skin infections.

- **gentle**
  designating an even-tempered, quiet, docile animal.

- **genus**
  a collection of species. One of the species is chosen as being representative and referred to as the genotype.

- **geophilic**
  occurring naturally in the soil. Used to refer to *geophilic fungi* that art found in the soil.

- **georgic**
  relating to agriculture.

- **gerber test**
  a test used by the milk marketing board to determine the butter fat content of milk. It involves the use of sulphuric acid and amyl alcohol to break down the proteins which surround the fat globules in milk releasing the fat for measurement.

- **germ**
  1. a tiny organism, bacterium, or microbe that can cause disease in man and/or animals.
  2. early embryo
  3. seed.

- **germ free**
  designating an animal that is free of harmful organisms, as germ free pigs used for

experimental purposes.

- **goose**
  a big bird of the duck family (*Anatidae*). Domestic geese are descended from the greying goose and are mostly kept m small flocks under free-range, providing table birds. Some large commercial units exist. The term goose is usually applied to the adult female.

- **goose stepping**
  in the pig this can be a symptom of a deficiency of pantothenic acid, one of the B group of vitamins, in cattle, it may be a symptom of familiar ataxia.

- **gosling**
  any young goose before its sex can be determined.

- **gossypol**
  a toxic yellow pigment found in cottonseed, heat and pressure tend to bind it with protein and thereby render it safe for animal consumption. it may cause discolouration of egg yolks during cold storage.

- **grade**
  an animal that is not a purebred but commonly showing characterises of a particular breed. usually one of their parents was purebred.

- **grade a milk**
  milk produced and processed under rigid sanitary regulations. milking and processing facilities are approved and inspected by public health officials, fluid milk products must meet this inspection standard.

- **grade cattle**
  animals possessing the distinctive characteristics of a particular breed but not registered with a breed association. most grades have one registered parent, usually the sire.

- **grading**
  separating the unopened fleeces into piles according to fineness and length.

- **grading up**
  the establishing of a pedigree herd of cattle by mating pedigree bulls with a core of non-pedigree cows, and subsequently mating each generation of females with the pedigree bulls or the same breed, thereby increasing the percentage of pedigree 'blood' with each generation.

■ **gram calorie**
represents the quantity of heat required to raise the temperature of 1 gram of water at 15° c by 1°c and is denoted shortly by the word *'cal'* written with a small 'c'. The gram calorie is a very small unit of heat, the kilogram calorie denoted shortly by the word *'Cal'* written with a capital 'c' is employed a Kilogram calorie is the quantity of heat required to raise the temperature of 1000 gm. of water at 15° c by 1°c and is, therefore, equal to 1000 gm. calories.

■ **gram negative bacteria**
bacteria that are decolourised by acetone or alcohol and therefore have a pink appearance when counterstained and safranine, e.g. *escherichia coli, salmonelia, vibrio* and *pasteurella*.

■ **gram positive bacteria**
bacteria capable to retain a crystal-violet dye even when exposed to alcohol or acetone e.g. staphylococci, streptococci and *Bacillus anthracis*.

■ **grand period of growth**
the period of growth of cells, tissue, organ, or organism. This starts slowly immediately after differentiation, increases to a maximum rapidly and finally falls off to zero.

■ **granulations**
are small masses of cells of a productive nature containing loops of newly formed blood vessels that spring up over the surfaces of healing wounds.

■ **granuloma**
is a tumour or other abnormal formation made up of granulation tissue which surrounds a central point of irritation.

■ **granulosa**
cells coating the ovarian follicle.

■ **GRAS (generally recognised as safe)**
food additives that have judged as safe for human consumption by a panel of expert pharmacologists and toxicologists who use data available, including experience of common use in food. The common use fac-

tor is often the major criterion on which judgement is based.

**grass**

any plant of the family generally in grassland agriculture, the term 'grass' does not include the cereals generally harvested for grain, but does include the forage species of legumes, often grown in association with grasses.

**grass juice factor**

jointly, the unidentified substances of a forage plant that contribute to the health and nutrition of the animal.

**grass land agriculture**

a scheme of farming in which either forage crops or soil conserving crops, usually both, forms the nucleus.

**grass silage**

a term designating fodder from grasses, legumes, and mixtures of them. Terms such as 'alfalfa silage', 'clover silage', 'alfalfa-timothy silage', are definite expressions which should be used where possible.

**grass tetany (hypermagnesemia)**

a form in ruminants characterised by a low level of magnesium in the blood. sometimes observed soon after recently freshened cows are turned on lush pasture. It is believed to be caused by an imbalance, resulting from precipitation of magnesium in the intestine, through large amounts of ammonia formed from nonprotein nitrogen in the forage.

**grass waterway**

a exterior drainage-way, usually broad and somewhat flat, protected from erosion by a natural or seeded grass cover.

**grassland**

any plant community in which grasses and/or legumes comprise the dominant vegetation.

**gravid**

pregnant.

**graze**

to consume standing vegetation, as by livestock or wild animals.

■ **grazier**
one who pastures livestock, ordinarily for commercial purposes.

■ **grazing, alternate**
the grazing of two or more pastures into succession.

■ **grease wool or greasy wool**
wool that has not been washed or worn.

■ **green bone**
bone cut fresh from the carcass at the butcher's the term implies freshness and has nothing to do with the colour.

■ **green cheese**
a term applied to the fresh uncured product. It refers to flavour, odour, body and texture, not colour.

■ **green chop (fresh forage)**
forages harvested (cut and chopped) m the field and hauled to livestock. This minimises the loss of colour, moisture, wastages and nutrients. Also called zero grazing.

■ **gut**
those parts of the alimentary canal having a primary digestive function .

■ **gut tie**
the informal name for a type of hernia in which a piece of bowel becomes intertwined in the spermatic cord following castration of cattle.

■ **guttate**
characterised by lesion that are drop-shaped.

■ **h antigen**
a type of antigen found in the flagella of certain members of the salmonella group of bacteria.

■ **habitat**
1. Kind of a place in which a plant or animal lives such as forest habitat, grassland habitat etc.
2. Often used for the natural range of distribution of a species.
3. Sum of effective environmental conditions under which an organism lives.

■ **habitat factor**
the influence of the habitat on the growth of an organism.

■ **habits**
all the acquired ways of feeling, thinking and doing which tend to repeat them-

selves under appropriate conditions.

- **habituation**
the gradual adaptation to a the environment or to stimulus. The extermination of a conditioned reflex by repetition of the conditioned stimulus (also called *negative adaptation)*.

- **hack**
1. a horse kept for hiring out. Especially an old or poor one.
2. an ordinary riding horse.

- **hackamore**
a bristle (with no bits) that controls a horse by pressure on its nose.

- **hackles**
the narrow lance – shaped feathers on the necks of fowls and saddles of cocks , i.e., saddle hackle and neck hackle.

- **hackney**
a breed of light draught horse, usually bay, brown or chestnut coloured.

- **haemangioma**
a tumour composed of blood-vessels.

- **haematemesis**
vomiting blood.

- **haematridrosis and haemotopedesis**
the presence of blood in the sweat.

- **haematocele**
a hole containing blood.

- **haematuria**
any state in which blood is found in the urine.

- **haemoglobin**
a complex organic compound containing iron and gives the red colour to the red blood corpuscles. It has the function of absorbing oxygen from the air in the lungs and of transporting oxygen to the tissues.

- **hagedoorn system**
a system of breeding where selection is made on sire family merit only.

- **hairpin loops**
areas of double helix formed by the pairing of the contigenous complementary stretches of bases on the same single RNA or DNA strand.

- **hairy fibres**
a term applied to coarse, straight, usually chalky fibres. In the Merino, they are more normally observed in

the breech region and on the folds of the neck.

- **half breed**

any animal of mixed breed. Specifically applied to a cross–bred. (Cross Breeding) type of sheep resulting from a cheviot ewe crossed with a Border Leicester ram.

- **half life**

the time in which the radioactivity originally associated with an isotope will be reduced by one – half through radioactive decay. The time in which half the atoms of a particular radioactive substance disintegrate to another nuclear form. Measured half-lives vary from millionths of a second to billions of years. Effectively Half Life is the time required for a radio nuclide contained in a biological system, such as a man or an animal, to reduce its activity by one-half as a result of radioactive decay and biological elimination.

- **half sib**

one of a pair of animals with a common parent, meaning a half-brother or half-sister.

- **half warp**

a wool having all the features of a warp wool, but less than the minimum length requirements.

- **halogens**

a group of elements, including fluorine, chlorines, bromine and iodine, which are strong oxidising agents and have disinfectant properties.

- **halxon**

an organo-phosphorus compound used against gastro-intestinal nematode worms.

- **halter**

a leather, canvas or rope device fixed firmly to the heads of cattle and horses for holding and leading them.

- **ham**

the thigh of a hog, prepared for food. The hind leg of a swine from the hock upwards on the live animal.

- **ham shank**

the hock end of a ham.

- **hammel**

1. a ran lamb.
2. a small shed.

■ **hamstring**
the huge tendon that is above and behind the hock in the rear leg of quadrupeds.

■ **hand**
term used to describe the height of a horse. It is the average width of a man's hand (4 in).

■ **hand milking**
the manual milking of an animal as opposed to machine or mechanical milking.

■ **hand feed**
feeding that is the result of the operator regulating the feed of each animal individually.

■ **handle**
a term which refers to the degrees of softness of wool to the touch.

■ **handlers**
dealers or processors of milk who commonly purchase *raw milk* and sell pasteurised milk and milk products.

■ **handling**
the individual manual examination of poultry to determine quality, health, etc.

■ **hangers**
the shorter sickle-like feathers at the base of the cock's tail; also termed tail coverts.

■ **hanging**
the practice of hanging animal and bird carcases, particularly game, on hooks at room temperature for a period so that the meat condition improves.

■ **hank (English worsted)**
560 yards of worsted yarn. The number of these standard hanks which together weigh = 1 pound specifies the count or size of the yarn.

■ **haploid**
state found in the sex cells or gametes in which half the usual (diploid) number of chromosomes are found. Condition necessitated by sexual reproduction and occurs during meiosis.

■ **haptene**
a substance that becomes antingenic when it combines with proteins.

■ **hard breeder (shy breeder)**
a female animal that has difficulty or is slow in conceiving.

**hard feathered**
feathers having a hard glued down appearance, or a variety of fowls carrying very scanty fluff.

**hard keeper**
an animal that is unthrifty and grows or fattens slowly regardless of the quantity or quality of feed.

**harden off**
to gradually reduce the temperature in a brooder over a period of time until young chicks no longer need artificial heat to survive. Nowadays temperature is usually reduced by 5.5° (10°F) per week over the first four weeks of life, from about 32° C to about 15.5° C (30° F to 60°F).

**hardy**
a term applied to plants and animals capable of surviving the cold during winter.

**harsh**
a term which describes wool lacking in softness to the touch.

**harvest year**
the first, second, third, etc., year following the seeding year of leys.

**hassock**
a clump or tussock of tightly packed grasses, reeds or rushes, etc.

**hatch**
to bring forth young from the egg by natural or artificial incubation.

**hatchability**
1. quality of a fertilised egg to produce normal embryonic development and normal emergence of the young, when incubated.
2. the percentage of fertile eggs that hatch.

**hatchel**
1. a Haycock
2. a comb for Flax or hemp.

**hatchery**
1. to either a cabinet or a walk-in type of incubator, or
2. to premises specialising in the hatching of eggs and production of Day-Old-Chicks, as distinct from the breeding or rearing poultry

**hatching**
the act of a young bird breaking out of an egg, aided by its egg tooth. A hen will normally sit on her eggs and incubate them to bring them to the point of hatch-

ing. Artificial incubation is used in commercial poultry rearing. A hatch is also a brood of chicks.

- **haversian**

discovered or described by Havers-the system of canals, ducts and lacunae within a bone.

- **hay**

a term mainly applied to grasses, but also to legumes, some herbs and occasionally cereals, which have been cut and dried and conserved for fodder. Hay may be classified into two main types, viz.,

1. seeds hay (sometimes called Clover Hay), usually derived from a 1 to 2 year old Leys, at one time mainly from a Red Clover-Ryegrass mixture and
2. meadow hay, derived from permanent Grassland or a long ley. Normally stored in bale form in barn nowadays.

- **hay bale stack**

bales of hay built into a heap for drying and storage, often under a Dutch Barn and generally on abase of rough timbers to raise the bales off the ground. Air spaces are left between the bales to allow moisture to escape and to allow lower bales to expand under the weight of higher ones. When chemical additives which prevent mould growth are introduced during bailing and hay is fully field-dried, bales may be more compactly stacked Small temporary stacks are usually left in the field for a period to ensure full drying before barn stacking. Field stacks are occasionally thatched with lay or reeds, but more commonly a plastic sheet is used to keep out rain.

- **hay quality**

treatment and weather conditions during haymaking can considerably affect hay quality and feeding value. Rapid drying is best, reducing cell respiration and carbohydrate loss by oxidation. Considerable carotene loss may occur due to excessive bleaching by the sun if the hay is left in the swath too long.

Protein and ash are more concentrated in leaves and are better conserved in hay harvested whilst still in leaty condition. Rain and dew can leach soluble nutrients from swaths and moulding, rotting and heating can occur in hay stored in too moist condition (i.e. above 25%)

- **hay rack**

a metal or wooden open frame in which hay is placed and from which livestock many take what they require.

- **hay rake**

a hand – rake with a long handle and normally with wooden prongs, used to gather hay.

- **hay stack**

a storage space heap of loose hay common before the introduction of the Baler and bale storage. Now generally only seen in hill areas and on small farms where topography precludes the use of a baler. Plastic sheets, as protection against rain, have now replaced thatch roofs on such stacks. Some stacks of baled hay are referred to as haystacks. Also called *hayrick*.

- **hay sweep**

a tool, either horse-propelled or fitted to the front of a tractor (and sometimes to an old car), once used commonly to pick up hay from Swaths or Haycocks and to transport it to a stack. It consisted of a series of wooden prongs about 3 m. (10ft) long and spaced at 30 cm. (1 ft) intervals in a frame. Often used in conjunction with an elevator of stationary baler. Nowadays most hay is collected using a pick-up-baler.

- **haycock**

a cone shaped heap of loose hay drying in a field. Now usually only found on hill farms and small holdings where the use of balers is not practical. Also called hatchel or pike. But a hayknife is a huge broad-bladed knife with a cross-set handle at one end, used to cut hay from a haystack, or fodder from a clamp.

- **haylage**

a fodder crop wilted in the field, after cutting to between 50-60% moisture content and then chopped and blown into an airtight Silo. Carbon dioxide released by cell respiration inhibits bacterial activity and rotting. Drilage is almost identical except that wilting is only allowed to between 60% and 70% moisture content.

- **hayloader**
  a trailed implement similar to a small elevator used to pick up loose hay from a Swath and deposit it onto a trailer for carting to haystack. Now seldom used except on some hill or small farms where the use of a baler is not practical.

- **haymaking**
  the conserving of hay for use as fodder, involving drying and the maintenance of nutritive content (hay quality). The moisture content of hay must be reduced from about 80% when freely cut (Mower), to below 20% to prevent moulding rotting, heating and possible combustion.

- **haysel**
  the haymaking season.

- **hayward**
  a person in the middle ages with the responsibility to fence and hedge repairs and preventing stock breaking through.

- **head**
  a hemispherical articular surface.

- **head shy**
  designating a horse on which it is difficult to put a bridle, to lead, or to work around its head.

- **head gear**
  the comb wattles and ear – lobes.

- **healds**
  frame suspended in the loom to which are attached wires that have eyelets in the centre through which the warp threads pass for making the pattern in the cloth.

- **healing powder**
  a delicately divided solid drug which is outwardly used for promoting the process of healing and protecting the wounds etc.

- **health**
  freedom from disease and pain. This refers to the nor-

mal condition of the body in which all its functions are normally performed.

- **health physics**
the study concerned with recognition, evaluation, and control of health hazards from ionising radiation.

- **heart**
the follow muscular organ which contracts rhythmically and maintains the circulation of the blood.

- **heart girth**
a measurement taken around the body just back of the shoulders of an animal, used to estimate body weight.

- **heat**
period of sexual receptivity in the female.

- **heat increment**
also called *work of digestion,* it is the increase in heat production following consumption of food when an animal is in a thermo neutral environment . It consists of increased heats of fermentation and of nutrient metabolism. There is also a slight expenditure of energy in masticating and digesting food. This heat is not wasted when the environmental temperature is below the critical temperature. This heat may then be used to keep the body warm.

- **heat labile**
unstable to heat

- **Heat of Fermentation (HF)**
the heat produced in the digestive tract as a result of microbial action.

- **Heat of Nutrient Metabolism (HNM)**
the heat produced as a result of the utilisation of absorbed nutrients.

- **heat period**
that period of time when a female will accept a male in the act of mating. Also called oestrous period in heat.

- **Heat Production (HP)**
results from
1. the heat used for maintenance (basal metabolism plus heat of voluntary activity) when an animal is consuming food in a thermo neutral environment. In direct calorimetry, heat pro-

duction is measured by use of an animal calorimeter.
2. the heat increment (heat of fermentation plus heat of nutrient metabolism) plus

- **heat stroke**
a condition caused by exposure of excessive heat; may occur in one of three forms:
1. sunstroke (thermic fever),
2. heat exhaustion,
3. heat cramps.

- **heat tolerance**
the ability of an animal to endure the impact of a hot environment without suffering ill effects.

- **heat treated milk**
milk subjected to heating, to destroy pathogenic bacteria and those organisms which cause souring. The method of heat treatment differs for the various designations of milk.

- **heavy breeds**
the sitting varieties.

- **hectare**
a metric unit of land measurement (2.47 acres)

- **heel**
back portion of the foot, which is composed of one single bone, the calcaneus (as calcis). The heel pad is made of fat and skin which bears the weight.

- **heel fly**
a name applied to the common cattle grub because it enters the body through the feet.

- **heifer**
a bovine female less than three years of ago who has not borne a calf. Young cows with their first calves are sometimes called *first calf heifers*.

- **heifer calf**
a female cow less than 1 year old. Also called *quey calf or cow calf*.

- **heiferette**
used to describe a heifer which has calved once, perhaps prematurely, then was 'dried up' and fed for slaughter.

- **heilman comb or rectilinear comb**
a type of comb used for combing short-fibred wools (French method).

- **helminth**
an wormlike parasite or intestinal worm.

- **haemangioma**
  a massing, over-growth or tumefaction of blood vessels.

- **haematic**
  relating to blood.

- **haematidrosis or sweating of blood**
  the mixing of blood with the sweat because of some circulatory disease.

- **haematin**
  an organic iron-compound which, in combination with a protein, forms the haemoglobin of the blood.

- **haematinic**
  an agent capable of improving the condition of the blood, especially the haematin content of the haemoglobin.

- **haematopoietic**
  an agent that promotes the formation of blood cells.

- **hemi cellulose**
  a group of complex carbohydrates relatively less resistant to digestion than cellulose.

- **haemo**
  a prefix used in many scientific terms. It means 'blood'.

- **haemoglobin**
  the red pigment in red blood cells of animals and man which carries oxygen from the lungs to other parts of the body. It is made of iron, carbon, hydrogen and oxygen and is essential to animal life.

- **haemoglobinemia**
  the presence of haemoglobin in the blood plasma.

- **haemoglobinuria**
  the presence of haemoglobin in urine.

- **haemolymph**
  the colourless blood of insects. Filling all the body cavities and combining the functions of blood and lymph.

- **ice milk**
  a frozen product resembling ice cream, except that it contain less fat (2 to 5 versus 10 percent) and more NMS (12 versus 10 percent) than ice cream. Both ice milk and ice cream contain a stabiliser or emulsifier and about 15 percent sugar.

- **ichtyosis**
  form of hereditary effect in calves characterised by

rough and cracked skin and huge, hairless patches over the body.

- **identical twins**

two similar individuals developed from the same fertilised egg. The egg separated into two parts shortly after fertilisation. If separation is late, Siamese (joined) twins may result (one individual with two heads or two bodies etc.). Identical twins are genetically alike. Therefore all differences between them represent environmental effect.

- **idiopathic**

a term applied to diseases with unknown cause.

- **idiosyncrasy**

a peculiarity of constitution causing an animal to react differently from most animals to a drug or treatment.

- **ileitis**

inflammation of the ileum.

- **ileocaecal**

the junction between ilium and caecum, that is between the end of the small intestine and the commencement of the large.

- **ileocecal valve**

valve at the junction of the lower end of thje small and large intestines (junction of ileum and caecum).

- **ileum**

the last portion of the small intestine which extends from the jejunum to the caecum.

- **iliac**

pertaining to the ilium.

- **illium**

the broad interior or upper part of the pelvis.

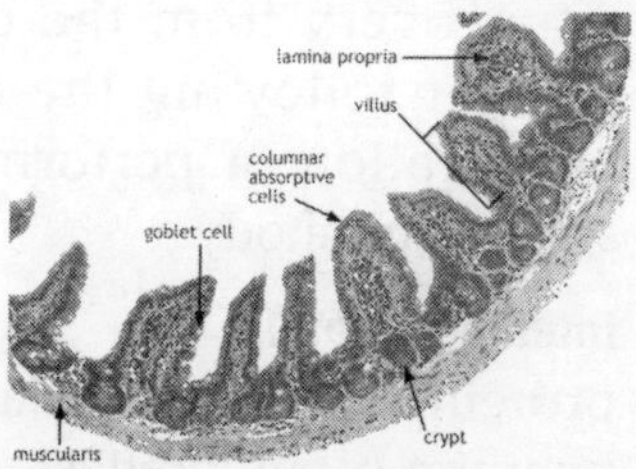


- **imbinbe**

to drink or inhale; to absorb moisture.

- **imino acid**

a secondary amino acid, with two hydrogen atoms from the ammonia replaced by alkaly radicals.

- **imitation milks**

mixtures of non-dairy ingredients (ingredients other than milk, milk fat and non-fat milk solids) which are combined, forming a product similar to milk, low fat milk, or skim milk. Sodium caseinate, although derived from milk, is commonly termed a non-dairy ingredient and is often used as a source of protein in imitation milks. Vegetable oils are commonly used as the source of fat.

- **immune**
a state of relative resistance to a specified disease following immunisation against, or recovery from the disease or following the administration or performed specific antibody.

- **immune globulin**
protein that contains large quantity of antibodies.

- **immunisation**
the process of artificially producing resistance to a given infection; generally by means of a vaccine, sometimes by means of an antiserum or antitoxin.

- **immunity**
the power of an animal to resist and/or overcome an infection to which most or many of its species are susceptible. Active immunity is attributable to the presence of antibodies formed by an animal in response to an antigenic stimulus. The administration of performed antibodies produces passive immunity.

- **immunise**
to render an animal resistant to disease by vaccination or inoculation.

- **immuno-globulins**
a family of proteins in body fluids which have the property of combining with antigen and, in the situation in which the antigen is pathogen, sometimes inactivating it and producing a state of immunity. Also called antibodies.

- **immunology**
the study of resistance to disease: the science dealing with the nature and cause of immunity from diseases.

- **impaction**
constipation.

- **impermeable**
that can't be penetrated

- **implant**
in radiology, radioactive material in a suitable container

to be embedded in a tissue for therapeutic purpose. It may be permanent (seed) or temporary (needles)

- **implantation**
  the process of attachment of embryos to the uterine wall.

- **import quotas**
  maximum legal imports permitted for various dairy products.

- **impotency**
  related to the capacity of the male to copulate.

- **imprinting**
  a mental process in which an inborn tendency in the animal causes it to attach itself to a set group of objects or a single object within a few hours after birth.

- **improved Dartmoor**
  a long–wool breed of sheep developed by crossing old moor land Dartmoor sheep with the Devon Long–wool breed. Found in localised areas only, mainly in semi-lowland situations.

- **in**
  a prefix used descriptively for the pregnant female of the species suggested by the added name of the appropriate young animal, which is being carried in the womb, e.g. in-calf, In-lamb. In-pig, etc.

- **in foal**
  designating a pregnant mare. Likewise, in calf refers to a pregnant cow.

- **in heat**
  refers to female animal at a period when she will accept coitus with a male.

- **in kneed**
  knock-kneed.

- **in lamb**
  a term applied to a ewe that is pregnant.

- **in lay**
  a hen during an egg laying period is said to be in lay.

- **in milk**
  designating a lactating female.

- **in season**
  a female animal in the state ready for matting. Also called *in-heat*

- **in situ**
  1. in the natural or usual place;

2. the usual site of origin.

- **in utero**
  within the uterus.

- **in vitro**
  in glass. Applied to biological experiments which are performed in test tubes or other laboratory glassware.

- **in vivo**
  within the living organism; applied to laboratory testing of agents within living organism.

- **inactivation**
  to destroy the activity of a substance, e.g.: heat treatment (56° C for 30 min) of serum abolishes complement activity in the serum.

- **inbreeding**
  production of offspring from parents more closely related than the average of a population. Genetically, inbreeding increases the proportion of homozygous genes in a population of 100 individuals that are all heterozygous (Aa), none would be homozygous (Aa), none would be homozygous. However, in a population in which 50 were homozygous (AA or aa) and 50 were heterozygous (Aa) the percent homozygous pairs of genes at this locus would be 50.

- **inbreeding coefficient**
  percentage which expresses the degree of extra homozygosity in an inbred animal due to the relationship between its parents.

- **incaparina**
  a so-called food substitute includes corn, cottonseed meal and torula yeast. First marketed in 1957, consumer acceptance has been slow.

- **inches of vacuum**
  measurement of degree of vacuum by measuring the distance the vacuum will draw a column of mercury up a tube.

- **incidence**
  the frequency of occurrence of a situation or a condition.

- **incisor**
  the name applied to the cutting teeth of each jaw. There are no upper incisor teeth in domesticated ruminants.

- **incomplete dominance**
  a state in which neither allele is dominant to the other, with the result both are expressed in the pheno-

type, which is intermediate between the two traits.

- **increment**
  the quantity or increase of growth in a given period of time.

- **incross**
  the result crossing of inbred lines within a breed.

- **incrossbred**
  the result of crossing of inbred lines of different breeds.

- **incrossing**
  crossing highly inred strains of the same breed.

- **incubation**
  1. the hatching of eggs, either naturally by a hen sitting of them, or artificially by keeping them in an artificially heated Incubator (Incubation Period).
  2. holding cultures of micro-organisms under conditions especially temperature, favourable for their growth. Specialised incubation permits control of humidity, light, radiation etc. in addition to temperature. Incubation is usually achieved with closed, head –insulated, thermostatically-controlled chambers. Sometimes, the term is also used to refer to the subjection of a specimen or culture to laboratory or a particular ambient for a given of time.

- **incubation period**
  1. the period between an animal becoming infected by disease causing germs (e.g., Bacteria) and the symptoms developing, during which germ numbers increase dramatically
  2. the period of time required for a newly laid egg to develop to the stage when the young bird is ready to hatch out. For chicken eggs the period is 21 days.

- **incubator**
  a kit used to hatch eggs by providing artificial heat, usually maintained at 37.2°C(99°F) and reduced for the last 2 days of the Incubation Period to about 36.1°C 97°F). Humidity is maintained at about 60%. Types vary from small flat ones holding 500-100 eggs, used for a small breeding flock, to large cabinet types holding up to 40000 eggs in trays or specially deigned

walk in rooms holding up to 80000 eggs, normally used at a Hatchery. Trays are frequently tilted to turn the eggs before they are transferred for the final 2 days of incubation to a hatching section.

- **independent assortment**
  a situation in which the separation or segregation of one pair of alleles has no effect on any other; occurs when the different pairs of alleles are on non homologous chromosomes.

- **independent culling levels**
  process of selection in which minimum satisfactory phenotypic levels are assigned to a number of traits independently and cows falling below any of these levels are culled regardless of their levels for the other traits.

- **index**
  a sum merit score, usually calculated for breeding animals. The weight of each component of an index is usually determined by its hereditary and economic importance.

- **indicator**
  a substance that changes colour as conditions change, e.g., pH indicators reflect changes in acidity or alkalinity.

- **indicator species**
  a species that has a narrow range of tolerance for one or more environmental factors so that, from occurrence at a site one can predict these factors at that site (e.g., nutrient availability or summer temperatures).

- **indigenous**
  an organism that is native to a particular habitat, as distinct from one introduced from outside the area.

- **indirect calorimetry**
  measurement of the amount of heat produced by a subject by determination of the amount of oxygen consumed and the quantity of nitrogen and carbon dioxide eliminated. The determination of heat production from the respiratory exchange.

- **indispensable amino acid**
  basically the same as essential amino acid.

- **inducible enzymes**
these arc enzymes whose rate of production can be accelerated by the presence of certain inducers in the cell.

- **inert**
relative inactive.

- **inert matter**
any material included in the ration which has no nutritive value. It may be used as a carrier for micronutrients.

- **infect**
to enter and establish a permanent or temporary parasitic relationship with an organism.

- **infection**
the invasion and presence of viable bacteria, viruses, parasites, etc., in a host which result in disease.

- **infectious**
capable of producing disease in a susceptible host.

- **infectious bovin rhinotracheitis**
an acute contagious vital disease affecting mainly the upper respiratory tract of cattle. Symptoms include runny eyes and nose, salivation. high temperature and loss of appetite together with a drop in milk yield in dairy cows. Fatalities are not common. The disease is more severe in store cattle.

- **infectious bronchitis**
acute, highly contagious vital disease of chicken which is characterised by respiratory symptoms like gasping and drastic reduction in egg production. It is caused by filterable virus having a particle size of 70- 106 u.

- **infectious diseases**
those diseases which can transmit from animals suffering to the disease free animals.

- **infectious equine encephalomyelitides**
characterised by symptoms of in coordination of movement and paraplegia. It has been caused by at least four distinct viruses which differ from each other in antigenic characters, mode of spread, virulence etc.

- **infectious laryngotracheitis**
highly contagious acute viral disease of chickens, characterised by gasping respiration rates, sneezing and coughing. It may be accompanied by expulsion of

blood stained mucus from the nostrils. It has been caused by a filterable virus which can pass through a pore diameter of 0.080u.

- **infectious necrotic hepatitis**
  acute infectious toxaemic disease which affects sheep and cattle. It is characterised by short course, black discolouration of the skin caused by subcutaneous sinuous congestion and necrotic foci in liver It is caused by localised bacterial infection of liver by *Clostridium oedematiens*.

- **infectious ovine encephalomyelitis**
  acute viral disease of sheep which affects horse, cattle pigs and man sometimes and is characterised by fever, peculiar louping gait, convulsions and paralysis. It is caused by Russian tick born complex of the group B arbor viruses.

- **infectiousness or inoculum**
  capability of inoculum to produce an infection in a susceptible host on arrival in infection court. It is reduced by resistance or immunity of host.

- **infective dose (id)**
  the number of micro-organ isms required or infect a host.

- **inferior**
  under-in human anatomy, posterior.

- **infertility**
  ability for reproduction is reduced but not eliminated.

- **intromission**
  the penetration of the female vagina by a male penis.

- **jack**
  an un-castrated male donkey.

- **Japanese B encephalitis**
  a virus which infects not only animals but also human beings. The disease is seasonal and the outbreaks are normally confined to the warmer months. The disease causes fever, loss of appetite, lethargy, sleepy appearance and jaundice.

- **jaundice**
  yellowish or orange discolouration of the tissues and retentions. It is especially visible in eyes, skin and mu-

cous membranes. It is caused due to circulation of bile pigment in the blood. Lever is mainly affected.

- **jejunum**
  the middle part of the small intestine which extends from the duodenum, to the ileum.

- **jenny**
  a female donkey (ass). Also called a *Jennet*.

- **jerky**
  long thin slice of sun-dried beef or lean meat.

- **jersey**
  a comparatively small, early maturing breed of dairy cattle imported from Jersey. Various coat colours exist from light fawn to darkish red, almost black, sometimes, with white markings. Characterised by a light coloured ring encircling the black muzzle, with a small head and dished face, Leanly built with a large well-developed udder. The average annual milk yield (Milk Yields) in 1980-81 was 3879 kg. and the rich creamy milk contains 5.22% butterfat and is much used for butter making.

- **jersian**
  a hybrid type of beef cattle derived from a Jersey bull crossed with a Friesian cow.

- **jetter**
  a way of connecting the liners and shells to the wash line for circulating cleaning.

- **joint III, joint felon**
  a disease of young livestock caused by various Bacteria entering the body via the unhardened navel and characterised by abscess formation at the navel and swellings limb joints. Also called navel-ill.

- **jowl**
  meat taken from the cheeks of a hog.

- **jugal**
  one of the facial bones of fowl.

- **jugular**
  the name of the large veins (one on either side of the neck) which carry the blood back to the chest from the head and interior parts of the neck.

■ **k selection**
a kind of natural selection which favours long lived, late maturing individuals. This species devotes small fraction of its resources into reproduction. For example tree species are K strategists.

■ **k value**
a measure of the insulating value of building materials such as glass fibre, wood.

■ **kahn test**
this test is used for diagnosis of syphilis in which *cardiolipin* (a non-treponemal) antigen is used. Such teats are also called Standard Serological *Tesas* (SSL).

■ **karyotype**
the chromosomes of an animal appeared at the metaphase of a somatic division; they can be arranged in pairs according to their length and location of the centromere.

■ **kebbing**
the enzootic absorption of sheep.

■ **ked**
a brown, hairy, wingless, flat, blood-sucking fly (*Melopoagus ovinus*), parasitic on sheep and causing constant irritation. Also called *cade, kade, kaid, sheep tick*.

■ **keel**
the breast bone; in exhibition ducks the pendant fold of skin hanging below the breast-bone.

■ **keep**
grass or other fodder crops on which livestock are grazed.

■ **kellner feeding standard**
the unit of measurement in this system is starch. Here one pound of digestible starch is taken as the net energy unit. The net energy value of any feeding stuff is expressed as the number of pounds of starch which would have the same net energy value for productive purposes. According to this system, a 500kg. animal needs 0.3 kg. of digestible protein and 3.175 kg. of starch equivalent. The starch equivalent of different feed ingredients are determined as follows:
Digestible protein X 0.94 =S E

Fat from coarse fodder 1.9=S E
Fat from cereal grainsx2.1=S E
Carbohydrates (digestible)+ fibre X1= S E
Fat from oilseedsx2.4=S E
This may be written as follows;
Digestible protein X 0.94 + digestible fat X 2.4+ nitrogen free extract + crude fibre-total starch equivalent.

▪ **kelp**

brown algae (seaweed) found in abundance in sea water. It is dried, ground or burnt for use as cattle feed. It is a good source of iodine, potassium and alginic acid. Some varieties grow to a length of 50 feet or more.

▪ **kemp**

an opaque fibre which is brittle and white almost like burnt bone. It will not take a dye. The presence of kemp constitutes a serious defect in wool.

▪ **keratin**

1. the protein substance which forms the wool fibre and is composed of the elements carbon (50.5 percent), hydrogen (6.8 per cent), nitrogen (16.8 per cent), oxygen (20.5 percent) and sulphur (5.4 per cent) approximately
2. a substance of which horn and the surface layers of the skin are composed. Basically it's a modified form of skin that has undergone compression and toughening.

▪ **keratitis**

the inflammation of the cornea, the clear part of the front of the eye.

▪ **keratoma**

a kind of horn tumour which causes deterioration the inner facet of the wall of the hoof.

▪ **keratosis**

a condition of excessive growth of horny tissue.

▪ **keto analogue**

a molecule which has the same composition as an amino acid except that there is an oxygen on the alpha-carbon instead of a hydrogen and an amino group.

▪ **ketone body**

any compound containing the carbonyl group CO.

**ketonurla**
the presence of ketone bodies in the urine, as with *ketosis* in high-producing cows.

**ketosis**
metabolic disease characterised by undue ketone body formation and high blood kelone level.

**key out**
the list of sources of variation and degrees of freedome of an experimental design.

**kid**
a young goat up to its first birthday.

**kidding**
parturition in the goat; the act of giving birth to young.

**killed virus**
a virus whose infectious capabilities have been destroyed by chemical or physical treatment.

**killing out percentage**
the weight of a dressed carcase as a percentage of the live weight of the animal (Deadweight). An sign of the amount of meat on a carcase after the removal of the head (except in pig), limbs, hide, blood and offal. For beef cattle the average is 54-57%.

**kilo**
a prefix that multiplies a basic unit by 1,000.

**kilocalorie**
1000 calories; the quantity of heat required to increase the temperature of one kilogram of water from 14.5°C to 15.5°C. Also called *large calorie*.

**kilopascals**
measurement of degree of cacuum in metric units. Written as kPa.

**kip**
a hide taken from a very young calf.

**kjeldahl**
relating to a process of determining the quantity of nitrogen in an organic compound. The amount of nitrogen measured is then multiplied by 6.25 to calculate the protein content of the food or compound analysed. This is a the method developed by the

Danish chemist JGC Kjeldahl, in 1883.

- **kline test**

a microscopic flocculation test for the diagnosis of syphilis.

- **knacker**

a person who slaughters animal.

- **knock kneed**

a deformed condition in which the legs bend in towards one another at the hocks.

- **kohl rabi**

a fodder crop of the cabbage group with a swollen, scarred stem, turnip-shaped, from which long-stemmed leaves are grown.

- **ksloes**

a black small, shaggy haired breed of cattle once found on the Hebridean islands and which contributed, with mainland cattle, to the development of Highland cattle. Sometimes used to mean *Highland cattle*.

- **kurtosis**

the departure of a symmetrical frequency distribution from the normal by deficiency (*teptokurtosis*) or excess (*platykurtosis*) in its shoulders, as opposed to tail and centre.

- **kwashiorkar**

a syndrome produced by a severe protein deficiency, with characteristic changes in pigmentation of the hair and skin, oedema, anaemia and apathy.

- **label**

an exhibition of written, printed or graphic matter upon the immediate container of any article. Pamphlets and other written or printed matter accompanying an article held for sale are also considered a part of the label.

- **labial**

lip-shaped organ, or lips.

- **labile**

unsettled or uneven. Readily or continuously undergoing chemical, physical, or biological change or breakdown.

- **lace**

a lighter or darker coloured border round the edges of some feathers.

- **lachrymation**

the act of tearing, secreting

and conveying tears.

- **lacing**
  a border, stripe of edge round a feather of a different shads or colour from that of ground, for example, in laced Wyandottes.
- **lacrimal**
  pertaining to tears.
- **lacrimation**
  the secretion of tears.
- **a-laetalburmin**
  a milk protein; also a subunit of the enzyme lactose synthetase.
- **lactate**
  an enzyme present in intestinal juice which acts on lactose to produce glucose and galactose.
- **lactate**
  to secrete or produce milk.
- **lactation**
  the period during which a female animal secrets milk from the mammary glands. In cows and goats kept for milk production the period is greatly increased by regular milking, without which secretion ceases. The optimal lactation for dairy cows is 10 months (305 days) from calving to when twice dairy milking ceases, thus allowing a 2-month rest (dry period) prior to calving again. Milk yield during lactation vanes with a peak at about 10-12 weeks after calving. Various factors affect the yield and quality of milk including breed, diet management and length of dry period.
- **lactation period**
  the number of days an animal secretes milk following each parturition.
- **lacteal**
  lymphatic vessel of the intestines that absorbs large amounts of fat, or chyle.
- **lactic acid**
  an organic acid, one form ($CH_3CHOH.COOH$) of which is normally found in sour milk, sauerkraut and silage. Other forms enter into body metabolism.
- **lactoflavin**
  term occasionally used for riboflavin or vitamin B. It is a water-soluble, yellow pigment having a greenish fluorescence. It is found in milk whey, egg yolk, egg white, leaves of plains and animal tissues.

- **lactogen**
  prolactin.
- **lactogenic**
  stimulating the secretion of milk.
- **ß-lacto globulin**
  a milk protein.
- **lactometer**
  tool used for testing the purity of milk. The scale on the stem has been marked M, 3, 2, 1 and W from the bottom upwards. Milk being heavier than water, the instrument sinks less in milk than in water. In pure water it sinks to the mark W but in pure milk it sinks up lo the mark M. It is the modified form of common hydrometer.
- **lactose**
  a disaccharide sugar (Carbohydrate) found in milk. A compound of glucose and galactose molecules, into which it is split by the enzyme lactase contained in intestinal juice.
- **lactose intolerance**
  a state in which there are clinical signs (abdominal pain, diarrhoea, bloating, flatulence) over and over again following a few hours after the intake of milk. Lactose is not sufficiently hydrolysed in the intestine because of insufficient quantities of the enzyme lactase and the high osmotic pressure that results causes infiltration of fluids into the intestines. Gas-producing micro-organisms of the lower intestine hay utilise lactose to produce gas.
- **lactose synthetase**
  two unit enzyme, composed, of galactosyl transferase and á-lactabumin. Catalyses the formation of lactose from glucose and UDP-galactose.
- **lairage**
  a place where live stocks are housed, mainly at markets and docks while awaiting slaughter for export.
- **lamarckism**
  the doctrine of Lamarck pertaining to inheritance of acquired characters.
- **lamb**
  a sheep less than twelve months old. To give birth to a lamb.
- **lamb hog**
  a male lamb from weaning time until it is shorn.

- **lamb marking**
  the act of ear-marking, tailing and castrating lambs.

- **lamb's wool**
  wool from sheep less than eight months old. It is soft and generally somewhat lacking in felling properties.

- **lambing percentage**
  the number of lambs born to 100 ewes.

- **lamella**
  a thin coating of substances surrounding an air cell such as the air cell membrane in frozen deserts.

- **lameness**
  a departure from the usual gait, occasioned by disease or injury situated in some part of the trunk or limbs and is usually accompanied by pain.

- **laminitis**
  inflammation of the lamina in an animal hoof, particularly cattle and horses, thought to be an allergic reaction lo overfeeding with barley, or to toxic products from afterbirth breakdown or bacteria.

- **lampas**
  a inflammation of the mucous membrane of the hard palate of the horse immediately behind the arch of the incisor teeth in the upper jaw.

- **lanolin**
  wool wax having greater purity than the ordinary wool grease purity than the ordinary wool grease of common degrees; the USP and cosmetic grades of lanolin have been highly purified, refined, bleached and deodorised.

- **lanugo**
  furry covering on a foetus, begins to shed before birth.

- **laparotomy**
  surgical opening of the abdominal cavity.

- **larva (plural is larvae)**
  the immature form of insects and other small animals, which is unlike the parent or parents and which must undergo substantial change of form and growth before reaching the adult stage caterpillars, maggots and screw worms are also larvae.

- **larvicide**
  a chemical used to kill parasite larvae.

- **latviparous**
  bearing and bringing forth young that are larvae (especially of insects).

- **larynx**
  the organ situated between the trachea and the base of the tongue, into which the glottis opens and which contains the vocal folds. A larynx is typically found in all lurked vertebrates above fishes, except birds.

- **late hatched**
  hatched out after May.

- **latent**
  hidden or not apparent.

- **latent period**
  pre-infective period.

- **latent period (radiology)**
  interval between irradiation and appearance of the effect in question.

- **latent tissue injury (radiology)**
  wound which does not expose until some time after irradiation and possibly until some other trauma has supervened. This second wound may be slight that it would not, or itself, produce observable damage.

- **latent virus**
  the virus not inducing development of symptoms in its host.

- **lateral**
  external -toward the outside - sideways.

- **Latin square**
  an experimental devise for comparing treatments where the number of replications is the same as the number of treatments and each treatment occurs only once in each row and column, especially adapted for accurate comparisons with a small number of treatments.

- **lavage**
  washing out of an organ, e.g., the stomach.

- **law of diminishing returns**
  law which tells an increase in the capital and labour applied to the cultivation of land causes in general a less than proportionate increase in the amount of produce unless it happens to coincide within improvement in the art of agriculture.

**law of segregation**

law stating that in eacn individual the genes occur in pairs and in the formation of gametes the members of each of pair separate and pass into different gametes, so that each gamete has only one each type of gene.

**law of statistical regularity**

this law states that a moderately large number of items chosen at random from among a very large group are almost sure, on the average, to have the characteristics of the large group.

**laxatives**

mildest and least active cathartic which is used to overcome constipation e.g., white mineral oil, sulphur etc.

**lay out**

the placement of experimental treatments on the experimental site where it may be over space, time or type of material.

**laying**

the expulsion of an egg. The term is commonly associated with hens in active egg production.

**laying birds**

most commercial hybrid birds produced from inbred strains. Light types are derived from the White Leghorn. Medium to heavy types are often derived from the Rhode Island Red. Other breeds used in hybrid production include Rhode Island White. Barred and White Plymouth. Rock, New Hampshire Red and Light Sussex. A synthetic strain is also used, derived from several breeds. There is limited demand for pure strains of Rhode Island Red, Ligh Sussex and Marans.

**laying period**

the period during which poultry are in lay. It is normally of about 50 weeks duration and usually commencing at 22-24 weeks of age far egg-producing strains and 26-28 weeks for broiller strains. Most egg-producing fowls are slaughtered after the first laying period as fewer eggs are laid in the second period.

**layer**

mature female fowl that has been kept for egg-laying purpose, especially one in current egg production.

- **lea**
open country such as meadow, pasture, or arable land left Fallow or under grass. Another term for ley.

- **$LD_{50}$ value**
a statistical estimation of the number of milligram of a given poison per kg. of body weight required to kill 50 percent of a large population of test animals. The LD value of a compound may refer to oral administration or to application to the skin.

- **lead feeding**
the practice of increasing grain led cows to a level equal to 1.0 to 1.5 per cent of body weight beginning about 3 weeks prior to the predicted calving date. Following parturition, grain is increased until a cow reaches maximum production or maximum feed intake, whichever, comes first.

- **leader**
the single spine at the rear of a rose comb, for example, on wyandottes.

- **leaf comb**
comb shape of butterfly with wings open; seen to advantage on Houdan cock.

- **leaky quarter**
the loss of milk through teat leakage.

- **learner (in shearing)**
a Shearer or intending Shearer who has not shorn 5000 sheep.

- **leas square estimator**
an estimator obtained by the method of least square.

- **least square method**
a technique of estimation by which the quantities under estimate are determined by minimising a certain quadratic form in the observation and those quantities.

- **lecithins (*Phosphatidyl Choline*)**
lipids containing phosphoric acid and chlorine in addition to glycerol and fatty acids.

- **leg**
the lower part of the shank which is covered with scales. Does not include the thigh.

- **leg feathers**
the plumage found on the outer side of the leg of such breeds as Brahmas and Faverolles.

- **leg bar**
an Auto breed of poultry

produced by crossing the brown Leghorn and the barred Rock. Barred in colour with yellow legs , a single comb and producing white eggs.

▪ **leggy**
too long on the leg.

▪ **leghorn**
a nervous, small, laying breed of poultry, variously coloured including blue, black, brown, black and white. Gold and silver and white, the last being the most popular yellow-legged, single-combed and producing large numbers of white egg.

▪ **legume**
a plant member of the leguminosae family, with the character of forming nitrogen-fixing nodules on its roots, in this way making possible in the use of atmospheric nitrogen.

▪ **lentigo**
a lentil -shaped, pigmented spot like a freckle.

▪ **leptotene**
the early phase of meiosis. The chromosome appear longitudinally, single rather than double and the structure is more definite than in mitosis. There is a series of dense cromosomes.

▪ **lesion**
injury or diseased condition reflected in discontinuity of tissues or organs, often causing loss of function of a part.

▪ **let down stimulus**
the stimulus a cow receives to induce her to release her milk.

▪ **leteolytic**
pertaining to the damage of the corpus luteum.

▪ **lethal dose (le)**
the number of pathogenic micro-organism required to cause death in a given species of plant or animal.

▪ **lethal gene**
a gene, or genes, causing the death of an individual. Most lethal genes are recessive or partially dominant, requiring two genes of the same kind in a pair (homozygous) to cause death.

▪ **leukocyte**
colourless or white blood cell of which there are a number of types including

neutrophils, basophils, eosinophils, monocytes and lymphocytes.

**leukoplakia**
flat, white patches of hyperkerantinisation on mucous membranes.

**ley corn**
a cereal crop which follow a Ley in a relation.

**ley farming**
a system of farming in which several fields (and sometimes an entire farm) are cropped as Leys with regular reseeding, or alternatively any rotation or cropping system which includes leys. Also called *alternate husbandry*.

**libe**
the term applied to the larger divisions of different organs, such as the lungs, liver and brain.

**libido**
sex drive or sexual desire.

**lice**
small non-flying sucking insects that are true parasites of men, animals and birds.

**lichenification**
a hardening and thickening of the skin characterised by an exaggeration of the superficial skin markings and usually accompanied by hyper-pigmentation.

**life cycle**
the changes in mode and form of life which an organism goes through between recurrences of the same primary stage.

**life history**
the continuous story of the actions and duration of each stage in the complete life cycle of a species during one generation or annual series of generation.

**ligament**
bands, sheets or cords of white fibrous or yellow elastic connective tissue which support or bind together parts of organs.

**light breeds**
breeds which do not sit. Mainly of the Mediterranean Class.

**light Sussex**
a variety of Sussex poultry characterised by white feathers and black marks on

the neck, wing tips and tail. A dual-purpose breed producing good quality white meat and brown eggs.

- **lignifications**
  the deposition of a number of substances called *lignin*, in the cell wall, which become intimately mixed with both the middle lamella and cellulose layers of the cell wall.

- **lignin**
  an hard to digest compound which along with cellulose is a major component of the cell wall of certain plant materials such as wood, hulls, straws and over-ripe hays.

- **limby**
  the length of the thigh in the Malay.

- **limestone ground**
  ground of calcium carbonate.

- **limiting amino acid**
  the indispensable amino acid of a protein which shows the greatest percentage deficit in comparison with the amino acids contained in the same quantity of another protein selected as a standard.

- **limiting essential amino acid**
  the amino acid present in a protein in the least amount in relation to the animal's need for that particular amino acid is referred to as the' limiting amino acid. Other essential amino acids can be used by the animal towards meeting its essential amino acid requirement only to the extent that the so called limiting amino acid is present.

- **limited feeding**
  feeding animals to keep up weight and growth but not enough to fatten or increase production. Feeding animals less than they would like to eat.

- **limousin**
  a French breed of beef cattle newly imported into Britain. Red in colour with somewhat lighter tanned legs. Large bodied, rapid growing and noted for producing a carcase with a high meat to bone ratio.

- **Lincoln curly coat**
  a now extinct breed of pig with a curly white coat, lopped ears and a heavy face which contributed to the development of the chester white.

- **Lincoln long wool**
  the heaviest and largest breed of sheep in Britain. Known for its long shiny wool which grows over the forehead and falls over the eyes and is much used for Justre yarns and sheepskin rugs. A polled, white-faced breed, which has been widely exported.

- **lincoln red**
  a big breed of evenly deep-red coloured cattle developed in Lincolnshire and closely related to the short horn, with short, forward add downward curling horns and a short, broad head and with a long muscular body and short stocky legs. Formerly regarded as a dual-purpose breed, it is now popularly crossed with dairy breeds to produce beef calves.

- **line (linea)**
  a small ridge, of bone-a thin elongate region of tissue of different colour or composition.

- **line bleeding**
  restraint of selection and mating to a single line or only closely related lines of descent.

- **linear model procedure**
  a process for estimating the ability of transmitting genetic materials of bulls by directly or indirectly comparing their daughters activities in the same herds and seasons using complex matrix manipulations; occasionally referred to as the direct comparison procedure and also as Best Linear Unbiased Prediction (BLUP).

- **linear regression**
  the deterioration represented by a straight line.

- **line cross**
  a cross of two inbred lines.

- **linear action**
  the squeeze-release movement of the liner.

■ **link**
a measure of length equal to 7.92 in. A 1/100th part of a chain.

■ **linkage**
the alliance of two or more genes, usually near each other on the same chromosome, so that they tend to be inherited together. The characteristics passed on by such genes are said to be linked.

■ **linkage group**
all the genes which have loci on a particular chromosome.

■ **linked genes**
groups of genes that tend to be inherited together and that are presumed to have their loci on the same chromosome.

■ **linnaen system**
a binomial system of naming animals and plants developed by the Swedish naturalist, Carl Linnaeus, in which the first name represents the Genus and the second the Species, e.g., Hordeum sativum (barley).

■ **linoleic acid**
an 18-carbon un-saturated fatty acid having two double bonds. It reacts with glycerol to form linolein.

■ **linolein**
an unsaturated fat formed from the reaction of linolenic acid with glycerol.

■ **maceration**
the softening and disintegration of tissue by wetting.

■ **macro**
large or major. Unusual size or length. Usually used as a prefix.

■ **macro-conidium**
the larger of two types of conidia in fungi that bear large and small spores.

■ **macro elements**
macro- nutrients. Elements which are required and found in relatively large quantities as natural constituents of organisms or tissues; major elements, like nitrogen, phosphorus and potassium.

■ **macro-environment**
the environment due to the general, regional climate, traditionally measured some 4 ft. above the ground

and away from large obstructions.

- **macromolecules**

huge molecules with a few thousand to hundreds of millions molecular weights.

- **macrophages (the large phagocytes)**

large scavenger cells, the function of which is primarily phagocytosis and the destruction of many kinds of foreign particles. These cells are strategically located in the spleen, liver, bone marrow, small blood vessels and connective tissue. Together, they compose the reticuloendohelial system.

- **macroscopic**

visible to the unaided eye of a size i.e., visible without the aid of microscope.

- **macule**

a visible circumscribed change in colour of, or on in the skin that is not visibly raised above or depressed below the surrounding general level of the skin. *Patches* are macules larger than 1 cm. in diameter.

- **maedi visna**

a gradually progressive virus disease of goats and sheep, first reported and named in Iceland (Etymologically maedi means 'laboured breathing; visna' means wasting) in 1939 and now widespread symptoms include slight distress. Infected sheep will lag behind others, breathing is fast and shallow, becoming progressively more laboured until the animal fails and dies after 3-8 months. No vaccine or treatment is available. Lambs arc infected from ewes via milk and the disease also spreads readily by close contact between animals.

- **maggot**

a legless larva of definite two-winged flies of the order Diptera, with a very small head. The maggots of many flies strike livestock and crops, e.g.. Blow Fly, Crane Fly, Warble Fly, etc.

- **maid**

an ewe teg shown to be barren after being served by a ram.

- **maiden**

an un-bred animal.

**maiden ewe**

the term is commonly applied to ewes that have not had their first lamb. An ewe that has not been served by a ram.

- **maiden seeds**
  a ley which has not been grazed or mown.

- **maillard reaction**
  also known as the *browning reaction* because milk turns brown as high heat induces the reaction, Concomitantly, a caramelised flavour develops. It is the reaction of milk protein with lactose, notably with the epsilon-amino group of lysine. Amino compounds catalyse sugar dehydration, fragmentation and condensation.

- **maintenance**
  permanence of an accession in its original form through careful growing out and proper storage.

- **maintenance ration**
  the quantity of food which when supplied to an animal will keep it healthy and maintain a constant liveweight, with no loss or gain.

- **maintenance requirements**
  the portion of the energy of the fond used in the life process of the animal for keeping the animal warm and for movement of the body.

- **maize bran**
  the bran, which essentially comprises the husk (pericarp and testa), is the outer coating of the maize kernel (a grain or seed) with varying proportion of the starchy part of the germ (the rudimentary form from which a new organism is developed seed, bud, etc.).

- **maize germ cake**
  the cake obtained after extraction of oil from maize germ.

- **maize germ oil cake (meal)**
  solvent extraction of oil of maize germ cake.

- **maize gluten meal**
  the dried residue from maize after the removal of the larger part of the starch and bran by the process employed in the wet milling manufacture of maize starch. It contains approximately 60% protein.

- **maize grit**
  the fine to medium sized crushed maize grain with little or none of the bran or germ.

- **malaria**
  a febrile disease of animals caused by the presence of certain protozoan parasites in the blood.

- **malformation**
  any irregular growth of apart from the body.

- **malignant**
  destructive or virulent as applied to cancer.

- **malnutrition**
  any disorder of nutrition. Usually used to specify a state of insufficient nutrition.

- **malt culms, coombs**
  a by-product of the malting process, used as a feeding-stuff for dairy cows, sheep and horses, after thorough soaking.

- **malt sprouts**
  the product obtained by removing the growing buds of malted barley and the mixing together with malt hulls or any other parts of malt.

- **maltase**
  an enzyme which acts on maltose to produce glucose. Salivary maltase is in saliva and intestinal maltase is m intestinal juice.

- **malting**
  the method whereby barley or other grain soaked in water and allowed to partly germinate (during which stored Starch is broken down to Maltose by the enzyme Diastase) and is then kiln dried, after which the roots and shoots are separated (malt culms) and the remaining grain (the malt) is used in brewing. The mall is steeped in water and the maltose extracted into solution (the wort), which is run-off, leaving the residual grain, which is used as an animal feeding-stuff.

- **maltose**
  a disaccharide produced by the enzymatic hydrolysis to starch by diastase. Maltose is metabolised by a wide range of fungi and bacteria; a carbohydrate;

- **mamma**
  milk secreting organ of female mammals.

- **mammals**
  a group of animals characterised mainly by hair and the secretion of milk to feed the young. It includes all farm livestock, except poultry.

- **mammary glands**
  milk-producing glands which is a characteristic of female mammals. Milk production and secretion is controlled by various hormones. Located in the udder in case of cows.

- **mammary veins**
  veins which carry blood away from the udder. It is found just under the skin on the stomach of cattle.

- **mammilla**
  the nipple of the female breast.

- **mandibles**
  the upper and lower halves of the beak.

- **mane**
  long hair on the back of a horse's neck.

- **mange**
  a skin disease of animals caused by Mites which results in the hair dropping out in patches, or due to persistent itching, being rubbed away by the animal.

- **manger**
  a trough in which food for horses and cattle is placed, often located in a stable.

- **mangy**
  infected with a skin disease or parasite so that the skin is dry and scaly.

- **mania**
  a kind of madness characterised by wild excitement, violent tendencies etc,

- **manioc**
  a tropical plant (*Manihot esculenta*) producing a tuber with a high starch but low protein content, used as a cereal substitute in animal feedingstuffs. Also called *cassava* or *tapioca*.

- **manual**
  pertaining to or performed with the hand.

- **manufactured milk products**
  these include cheese, butter, evaporated whole milk, condensed whole and skim milks, whole milk powder, non-fat dry milk, ice cream, frozen deserts, aerated cream, frozen and plastic

creams and milk used in candy, soup, bakery products and feeds.

- **manufacturing grade milk**
milk produced under regulations less strict than those for grade A milk production. Acceptable milk contains not more than 3.000,000 bacteria per millilitre.

- **manufacturing milk**
milk for the manufacture of dairy products such as cheese, butter, powdered milk, ice cream and condensed milk.

- **manure**
excreta of animals, dung and urine (usually with some bedding).

- **manurial value**
nutrients in a feeding-stuff which an animal does not take in and pass out with the faeces so as to enrich the dung. Also called The residual value or fertiliser value of a feeding-stuff.

- **manus**
the hand-the forefoot.

- **map distance**
on a chromosome, the distance in crossover units between designated genes.

- **maran**
also used as *table birds,* it is a weighty breed of poultry with barred feathers, white legs, a single comb, producing dark brown eggs.

- **marasmus**
a progressive wasting and emaciation. Etymologically it cames from the Greek 'marasmos', meaning 'a dying away'. Enzootic marasmus is a condition of malnutrition in domestic animals due to a Imla deficiency of one or more trade especially cobalt and copper.

- **marbled**
applied to the ticked and laced feathers on the Pile Game cock's breast.

- **marbling**
the distribution of fat within the muscular tissue which gives the meat a white spotted appearance

- **marden layer**
a term for a young ley.

- **mare**
a mature female horse (usually more than four years of age).

▪ **Marek's disease**
a virus disease of poultry causing lameness, progressing to leg paralysis, drooped wings, neck twisting and greying of their iris in both eyes.

▪ **margin (spread)**
the difference between the purchase price and the selling price.

▪ **marked coat**
an animal's coat having two or more specific colours as distinct from a single colour.

▪ **market milk**
milk which meets conditions that qualify for use as fluid milk in an organised marketing area. Grade A milk meets this requirement.

▪ **marking**
marking the ears for the purpose of identification.

▪ **marrow**
the spongy material filling the cavities within cartilage bones.

▪ **marsupial**
one of a class of mammals characterised by the possession of an abdominal pocket in which the young are carried for some time after birth.

▪ **marsupilia**
pouched mammals, e.g., the kangaroo.

▪ **masculinity**
physical appearance resulting from well-developed secondary male sex characteristics.

▪ **mash**
a mixture of animal foods which has been crushed and stirred to a soft pulpy state with water (wet mash) or without it (dry mash).

▪ **mash feeds**
a mixture of non-processed and/or individually processed ingredients or combinations of body.

▪ **mashlum, maslin**
a mixture of oats barley and sometimes wheat, sown to provide grain for feeding to livestock. Occasionally beans or peas may be added to the mixture but ripening is uneven and

they ate not suited to being harvested by a combine harvester. Also called dredge or meslen.

- **mass spectrometer**
  an instrument for separation and measurement of isotopes by their mass.

- **mast**
  the fruit of beech, chestnut, oak and other trees, eaten by pigs as pannage.

- **mast cell**
  a large cell containing meta-chromatically staining cytoplasmic granules and thought to contain heparin and histamine.

- **mastectomy**
  taking away of the mammary glands.

- **mastication**
  the chewing of food.

- **mastitis**
  inflammation of a cow's udder, due to bacterial infection, resulting in the secretion of clots of milk and sometimes pus when the infection is actue. Also called *mammitis, garget and udder clap.*

- **mastocytosis**
  a state in which mast cells are unusually abundant.

- **mat**
  a closely interwoven or tangled mass of dead grass which develops when a pasture is under grazed, particularly on acid soils.

- **maternal**
  pertaining to the mother.

- **maternal impressions**
  disbelief that phenomena other than the physiology of the cow can affect the calf, e.g., black fence surrounding pregnant females yield black calves.

- **maternity barn**
  building set aside for pregnant animals in order to reduce to a minimum the danger of infections and complications during parturition.

- **mating**
  the act of joining rams with ewes.

- **mating group**
  a class of individual, haploid or diploid, within which mating is favoured at the expense of mating outside the group, by genetic or environmental

conditions characteristics of the group.

- **mating rate**
  the percentage of sows or gilts exhibiting oestrus that are actually mated or served.

- **ME**
  Metabolisable Energy,

- **meadow**
  firmly applied to a region of grassland used to produce hay or silage rather than for grazing. Also used loosely for rich waterside fields which are frequently flooded by river water, either naturally or via sluices and which are grazed (water meadows).

- **matron**
  a mare that has produced a foal.

- **Maturation**
  The process of becoming mature.

- **mature**
  the age after which there is little increase in the size of the animal and animal is capable of producing its first offspring.

- **mature equivalency (me)**
  age conversion formulas applied to the milk-production records of young cows to predict their expected mature milk production potential. Breed ME factors are used for comparative purposes in selecting and mating animals.

- **mature equivalent**
  standardisation of lactation records to the level of yield that would have been attained by each cow if she had been a mature cow and calved in the month of the year of highest calving frequency for her breed.

- **maverick**
  an unbranded animal, particularly a calf. It is also referred to a motherless raff in some areas.

- **maw**
  the stomach, mainly the fourth stomach of ruminants.

- **maxilla**
  the upper jaw-bone which carries the cheek teeth. Part of the roof of the mouth and the floor of the nasal passage is formed by it.

- **meadow fescue**
  a loosely tufted perennial grass (*Festuca prateiuis*) form-

ing large russocks, with narrow, bright green, hairless leaves, shiny on the underside and loose, green or purple panicles. Valuable for hay and grazing and particularly found on rich, well-drained, fertile clay soils and often abundant in water meadows. A very palatable grass which grows mixed with white clover or timothy.

- **meal**

1. another term fur oil seed cake.
2. a type of feeding-stuff consisting of a single or a mixture of finely ground ingredients, e.g., cereals, oil seeds, meat, bone, fish, etc. Mainly used for pigs and poultry.

- **meal feeds**

a ground ingredient, or ingredient otherwise reduced in particle size or sometimes referred to as combinations of ground ingredients.

- **mealy**

feathers stippled with a lighter shade giving the appearance of having been dusted with flour.

- **mean**

statistical and mathematical term for the average value of a set of values.

- **mean cycle length**

the average of the interval between any two consecutive heats of animals; when longer than normal, indicates embryonic death with the presence of venereal disease or undetected heat.

- **mean deviation**

the arithmetic average of deviations of all the values taken from a statistical average (mean, mode or median) of the series. In taking deviations of values, algebraic signs+ and - are not taken into consideration, that aside ration, that is negative deviations are also treated as positive deviations.

- **mean population**

the arithmetic mean of a variant for all members of a population.

- **mean square**

the variance; so called because it is the average of squared deviations from the arithmetic mean of a series of values.

■ **mean square error**
the amount whose square is equal to the sum of the squares of the individual error divided by the number of those errors.

■ **measles**
a condition of the carcases of cattle and pigs characterised by presence of tapeworm cysts in the meal.

■ **meat**
the flesh of animals used as food. It contains proteins, fats, minerals, vitamins and water. The most common meals used as food are lamb, beef and pork. Poultry and fish are sometimes included with the foods classified as meat.

■ **meat bird**
a fowl domesticated for its meat as contrasted to one kept for egg laying.

■ **meat meal**
the finely dried ground residue from mammal tissue exclusive of hair, hoof, horn, stomach contents and hide trimmings, except in such traces as may occur unavoidably in good factory practice.

■ **n**
1. it stands for a negative test in tuberculin test records.
2. number of chromosomes that make up one genome. It is called the basic, haploid, or monoploid number.

■ **NAD**
abbreviation of Nictinmide Adenine Dinucleotide, a coenzyme in metabolism which serves as a hydrogen and electron transfer agent by virtue of reversible oxidation and reduction. Formerly called as DPN.

■ **NADP**
Nicotinmide Adenine Dinucleotide Phosphate, a coenzyme in metabolism which serves as a hydrogen and electron transfer agent by virtue of reversible oxidation an reduction. Formerly called TPN.

■ **nag**
a horse or pony of nondescript breeding.

■ **nanism**
dwarf growth.

■ **nano**
one part a quantity when divided into 1 billion ($10^9$). Usually used as a prefix.

■ **nap**
ends of fibres raised from the body of the yarn, so as to the cloth of fibrous surface.

■ **nape**
back of the neck of an animal.

■ **naves**
is the Latin word for the nostrils.

■ **national plan**
a set of regulations for breeding, established to bring about improvement of poultry stocks on a national scale.

■ **native grass**
a grass species indigenous to a region; not introduced from another environment or area.

■ **native pastures**
a term referring to native vegetation in an unfilled prairie or plains area. The term 'unimproved' is used instead of 'native' for pastures which include mainly species not native.

■ **naturally antibody**
iso-agglutinin. An agglutinin that acts upon the red blood cells of members of the same species. Also called Iso-haemaglutinin.

■ **natural eco-system**
an eco-system that ultimately produces the climax community, which develops in the absence of any major disturbance and consists of a complex, self regulating food web of several Tropic levels.

■ **natural immunity**
immunity to a disease or infestation which results from qualities inherent in an animal.

Naturalradiation or background radiation.

Natural selection according to Darwin's theory of evolution, organisms tend to produce progeny far above the means of subsistence: a struggle for existence ensures which results in survival of those with favourable variations. Since the favourable variations accumulate as generations pass, the descendant tend to diverge markedly from their ancestors and to remain adapted to conditions under which they live.

- **natural service**
  in farm animals, it means to allow natural mating as opposed to artificial insemination.

- **naturalisation**
  the process of successful adaptation of plants or animals introduced to areas where they are not indigenous.

- **nausea**
  a stomach sickness characterised by a tendency of vomiting.

- **navitulai**
  the distal sesamoid of the horse's foot.

- **NDM (Non-fat Dry Milk)**
  the product obtained by removing water from pasteurised skim milk. It contains not more than 5 per cent moisture and not more than 1-5 per cent MF unless otherwise indicated.

- **neck**
  a constricted portion of a bone which connects the head to the shaft.

- **neck collar**
  a broad strap which fits round the neck of a cow or horse, securing it in a stall.

- **necropsy**
  an examination of the internal organs of a dead body to determine the apparent cause of death. Also called autopsy; fostmortfm.

- **necrosis**
  death of tissues, usually hi individual cells, groups of cells, or small localised areas.

- **needle teeth**
  eight small, sharp, tusk-like teeth, which pigs have at birth.

- **needles**
  sharp, slender, pointed, straight or curved instrument used for puncturing the skin etc.

- **NEg**
  net energy for body weight gain.

- **negative correlation**
  two variables are said to be negatively correlated if an

increase in one variable goes hand in hand with a decrease in the other.

- **negative feedback**
device whereby a hormone acts to inhibit the secretion of another hormone.

- **negative staining**
a type of staining wherein the background rather than the cells or tissues is stained or made opaque. It is used to demonstrate bacterial capsules.

- **negri bodies**
minute pathological structures (acido-philic, intracytoplasmic inclusion bodies) found in certain brain cells of animals infected with rabies virus. Negri bodies may be present only in the late stages of the disease. They may contain virions or viral components.

- **NEi**
net energy for lactating cows.

- **neigh**
the characteristic cry of a horse.

- **neisseria**
spherical. Gram-negative bacteria, some of which are associated with eye infections.

- **NEM**
net energy for body maintenance.

- **nematodirus disease**
a disease of lambs caused by parasitic worms (Nematodirus spp.). They are transmitted by eggs in the faeces which later hatch in pasture to infect the following year's lambs, causing diarrhoea and lack of thrift and some times anaemia.

- **neonatal**
pertaining to a newborn animal.

- **neonate**
a new born infant.

- **neoplasm**
literally it means a new growth and is applied to tumours in general.

- **nephrectomy**
the name given to the operation by which one of the kidneys is removed.

- **nephritis**
inflammation of the kidneys.

▪ **nephron**
the secretory portion of the kidney tubules.

▪ **neps**
small knots of tangled fibres removed in the combing process.

▪ **nerving**
the term is applied to serving the artery between the hoofs of animals.

▪ **nervour**
relating to nerves.

▪ **nervous system**
it consists of cell the nerves, nerve centre. nervous tissues and ganglia.

▪ **nest boxes**
boxes in which poultry kept under a system of deep litter are able to lay eggs. They may be of the single type designed to accommodate 5 birds, or the communal type, accommodating unto 100 birds. On large poultry farms the boxes are often arranged so that newly laid eggs roll on to a conveyor belt for delivery to a packing room.

▪ **net energy**
the energy from a feed after deducting from its total energy value the amounts lost in
1. faces,
2. combustible gases,
3. urine and
4. heat increment.

Net energy is commonly expressed in therms. A therm is the amount of heat required to raise the temperature of 1000 kilograms of 10C.

▪ **Net Energy for Maintenance (NEM)**
the portion of net energy expended to keep the animal in energy equilibrium. (There is no net loss or grain of energy in the body tissues).

▪ **Net Energy for Production (NEN)**
the portion of net energy required in addition to that needed for body maintenance which is used for work or for tissue gain (growth and/or fat production) or for the synthesis of productive and products (a foetus, milk; eggs, wool, fur, or feathers).

▪ **net protein ratio**
the difference between the average final body weight of

a test group of animals fed a protein diet and that of a control group receiving a protein free diet, divided by the amount of protein taken by the test group.

- **neural**
  pertaining to nerves.
- **neurectomy**
  an process in which part of a nerve is excited. The process is from time to time performed to give relief from permanent lameness in the horse but only a few months, work may be gamed.
- **neurilemma**
  the thin membranous cover of nerve fibres.
- **neuritic**
  pertaining to the nerves.
- **neurolibroma**
  a timorous lesion which is a mixture of abundant fibrous and nervous tissues.
- **neuro-hormone**
  hormone which is synthesised in neurons.
- **neuro-hypophysis**
  posterior pituitary.
- **neurone (nerve cell)**
  the primary cell type of the nervous system, which is specialised for the transmission of information. Neurones re arranged to form an integrated system of communication channels.
- **neurone theory**
  the theory that the nervous system is-composed of discrete cells called neurones, which do not physically join but become closely apposed at region called synapass, which are specialised for the transmission of information from cell to cell. This theory superseded the reticular hypothesis, which claimed that a protoplasmic contact existed nerve cells.
- **neurotoxiciiy**
  the condition of being poisonous to the brain and the nerves.
- **neurotransmitter**
  chemicals associated with the nerves which are involved in transmission of impulses.
- **neuter**
  an organism that does not have sex organs but is otherwise normal.

- **neutral fats**
  fats containing glycerol (alcohol esterifies with fatty acids).
- **neutron (n)**
  an uncharged elementary particle with a mass slightly greater than that of the portion, found in the nucleus of every atom heavier than hydrogen. Neutrons sustain the fission chain reaction in a nuclear reactor.
- **neutron activation analysis**
  activation analysis in which neutrons are the activating agent.
- **neutrophil**
  a type of white blood-cell which can migrate into the tissues and engulf bacteria etc.
- **nevus**
  a pigmented spot on the skin. However, was inference can be attached to the word. There arc many types, such as blue nevus compound nevus and junction nevus.
- **new breed**
  a breed not yet accepted to the Poultry Club Standards.
- **new process**
  pertains to the extraction of oil from seeds. Same as expeller process.
- **Newcastle disease**
  a extremely infectious noticeable disease of poultry caused by paramyxo virus. Symptoms include reduced yield of often, soft and misshaped eggs, lack of appetite, troubled breathing, nasal discharges and foul-smelling yellowy, watery diarrhoea. High mortality amongst young birds is common.
- **NFS**
  non-fat solids of milk. It is composed of proteins, lactose, minerals and other water soluble constituent of milk.
- **niacin**
  one of the B vitamins. Same as nicotinic acid. Nicotinamide also has niacin activity.
- **nick (niche)**
  the product of a certain mating which produces an animal of high order of excellence, sometimes from mediocre parents.

- **nicking**
  the making of progeny that are superior to the parents which produce them. It is sometimes referred to as *heterosis*.

- **nicotinamide**
  the amid of nicotinic acid- it has niacin activity.

- **nicotinic acid**
  same as niacin. One of the B vitamins.

- **nictitans gland**
  gland of the third eyelid (formerly called *harderian gland*).

- **night blindness**
  the incapability of animal to see well in dim light. It is the first symptoms of vitamin A deficiency.

- **ninhydrin reaction**
  blue colour reaction used in the chromatographic classification of amino acids. A blue compound is formed when ninhydrin is heated with alpha-ammo acids or with peptides or proteins which contain at least one free amino and carboxyl group.

- **nit**
  the egg of a louse or similar insect.

- **nitrate poisoning**
  a condition sometimes resulting when ruminants ingest nitrates ($NO_3$) which rumen bacteria convert to nitrite ($NO_2$)- The nitrites compete with oxygen, tying up the oxygen-carrying mechanism in the blood, causing the animal to suffocate.

- **nitrogen balance**
  the nitrogen in the food intake minus the nitrogen in the faeces minus the nitrogen in the urine. Nitrogen retention,

- **Nitrogen Free Extract (NFE)**
  comprises the carbohydrates, sugars, starches and a major portion of the material classed as hemicelluloses in feeds. When crude protein, fat, water, ash and fibre are added and the sum is subtracted from 100, the difference is the NFE.

- **nitrogenous**
  relating to or containing nitrogen.

- **NMS**
  non-fat milk solids (protein, lactose and minerals of milk). Also called NFS.

- **noble comb**
  a tool used in the English method of combing for the longer-fibred wools.

- **nocturnal**
  1. of the night;
  2. a nocturnal parasite is one that is active at night.

- **nodule**
  a tiny node or circumscribed solid elevation that usually extends into the deeper layers of the skin.

- **noil**
  the short fibres removed from the long ones in the processes of combing.

- **nomograph**
  a figure for the graphical solution of problems that involve formulas in two or more variables.

- **non-disjunction**
  failure of a pair of homologous (sister) chromosomes (dyads) to separate during the reduction division of meiosis. Thus both chromosomes go into a gamete (usually only one of each pairs goes into a gamete).

- **nonessential amino acid**
  any one of several amino acids (hat are required by animals but which can be synthesised in adequate amounts by an animal in its tissues from other amino acids.

- **non-protein**
  any one of a group of ammoniac nitrogen containing compounds which are not true proteins Urea is a common example.

- **Non-protein Nitrogen (NPN)**
  nitrogen in feed from substances such as urea, ammonium salts, amines, amino acids, but not from preformed protein.

- **non-return**
  the breeding efficiency of bull expressed as the percentage of cows that conceive on the first service.

- **non-ruminant**
  a simple-stomached animal that does Dot ruminate. Examples are swine, horses, dogs and humans.

- **non-septate**
  having no dividing cell wall in a filament.

- **non-sitter**
  varieties which due not incubate their eggs.

■ **normal distribution**
the limit which is reached cither by the binomial or the multinomial series, where tie power is large and. none of the summed quantities vary small in relation lo the power and to one another. This frequency distribution it, expected from a series of observations on a variant whose magnitude is affected by a large number of agents having small independent effects.

■ **normally**
concentration or strength of a chemical on the normal scale. A normal solution contains one gram atom of replaceable hydrogen or its equivalent per litre.

■ **nostril**
one of the external openings of the olfactory sac or nasal cavity of a vertebrate; an external naris.

■ **not sexed**
a day old chicks which are sold straight from the incubator as hatched and have not been sexed.

■ **notching**
cutting dents in the ears of animals for identification.

■ **noxious**
harmful, not wholesome.

■ **NPN**
Non-protein Nitrogen (e.g., urea).

■ **nubian**
a short-haired breed of goat which originated in North-East Africa and was crossed with short-haired British goats to produce the Anglo-nubian breed. Characterised by long dropping ears, a Roman noseband an arched face with prominent fore-head. Various colours exist, the most common being red, black and tan, often combined with white. Noted for its milk rich in butterfat.

■ **nuclear membrane**
a coating bounded to nucleus. It consists of two layers (each about 75-100A$^0$ thick) bounding a cisterna-like space and the outer one being continuous with the endoplasmic reticulum. It is perforated by round holes (200-500A$^0$ in diameter) called nuclear pores.

■ **nuclear reactor**
a device in which a fission chain reaction can be initiated, maintained and con-

trolled. Its essential component is a core with fissionable fuel.

- **nucleic acid**
one of a series of compounds in which phosphoric acid is combined with carbohydrates and with various bases derived from purine and pyrimidine. They are often conjugated with proteins to form nucleoproteins.

- **nuclein**
a protein substance containing phosphorus derived from the nuclei of cells.

- **nucleo cytoplasmic ratio**
the ratio of nucleus volume to cytoplasm volume. Alteration of this ratio may be a cause of cell division.

- **nuceloside**
a compound composed of a nitrogen base and a sugar.

- **nuceleoside**
a chemical compound composed of a nitrogen base, a sugar and a phosphoric acid molecule.

- **nucleus**
1. the small, positively charged core of an atom. All nuclei contain both protons and neutrons except the nucleus of ordinary hydrogen, which consists of a single proton.
2. a deep-staining body within a cell, usually the center the heart and brain of the cell, containing the chromosomes and genes.

- **nuclide**
a general term applicable to all atomic forms of the elements.

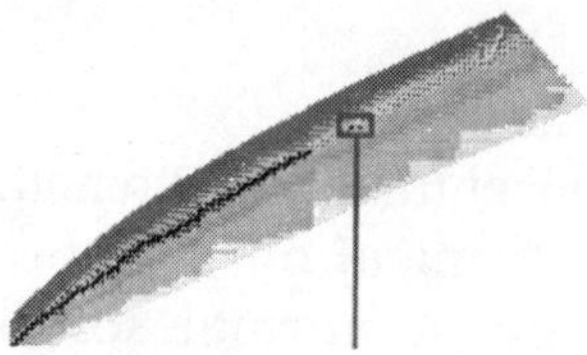

- **nulliparous**
having never given birth to viable young.

- **nurse**
to suckle; to give milk to (e.g. a baby).

- **nurse cow**
a milk cow used to supply milk for nursing calves other than her own.

- **nutrient**
a substance (element or ingredient) that nourishes the metabolic processes of the body. It is one of the many end products of digestion.

■ **nutrient animal**
any food constituent, or group of food constituents of the same general chemical composition, that contributes to the support of animal life.

■ **nutrient cycling**
the pathway of nutrient substances from their occurrence in the physical environment to their return lo the physical environment through the metabolic activity, death and decay of organisms.

■ **oat feed**
a kind of by-product consisting of fraction of oat grains, hulls and also dust. The product is obtained during milling of table cereals from clean oats The crude fibre percentage docs not exceed 25 %.

■ **oat groats**
these are oats grain from which the hull has been removed In other words, these are oat kernels. Since most of the feeding value of oats grain is found in the kernel, oat groats are very high in feeding value.

■ **oat hulls**
the outer covering of threshed oats obtained during the milling table cereals.

■ **oat meal**
an animal feeding-stuff consisting of ground oats with the husks removed to varying extents. The husks provide a roughage element to the diet. Fed either dry or together with other constituents in a wet mash. Also fed as a porridge (sometimes called *brose)* to animals which are ill or recovering from sickness.

■ **obese**
being overweight due to a surplus of body fat.

■ **obligate**
necessary or required: refers to any state or condition which is an essential attribute of a given organism. Thus, e.g., an obligate arobe grows only in the presence of free oxygen; an obligate parasite is one which grows only as a *parasite.*

■ **obligate parasite**
a parasite incapable of living without a host.

- **observation error**
the difference between an observed value of a quantity and a value adopted as representing the ideal true value of that quantity. Errors of observation are composed of either one or both of two general classes of errors: Accidental errors and systematic errors. The algebraic sign of an observation error is determined from the equation: Error=observed value—adopted ideal value.

- **obstetrics**
the art of the delivery of the young and the study of the abnormalities and diseases incidental thereto.

- **obturator**
a natural or man-made disc or plate closing an opening.

- **obturator foramen**
the hole in the pelvic formed by the junction of the ilium, ischium and pubis.

- **occiput**
the topmost posterior part of the head where it meets the neck. The occipital bone lies in the part of the skull which forms the occiput and can be felt as a hard bony plate in most animals.

- **occult**
difficult to understand; concealed from observation;

- **oddments**
butts, bag and fadges containing out-sorts from the classer's main lines.

- **oesophaotomy**
a surgical operation involving opening of the oesophagus for taking away of a foreign body, etc.

- **oestrogen**
hormone secreted from the early ovarian follicles.

- **oestrum**
the period during which an ewe will accept a ram's service.

- **oestrus detection**
the usual and rigorous observation of sows or gilts for signs of oestrus.

- **oestrus (heat)**
the showing of signs of sexual receptivity in the female under the influence of high levels of the hormone oestrogen in the blood stream.

- **oestrus induction**
an agent or method, used to encourage a sow or gilt to show oestrus at a time

when she would not generally have done so.

- **oestrous synchronisation**
any method which causes a group of females all to show oestrus at roughly the same time.

- **off feed**
having ceased eating with a healthy and normal appetite.

- **off feed**
refers to an animal that eats very Little or has stopped eating (usually the result of having eaten too highly concentrated feed or too great a quantity). Most often occurs with fattening animals.

- **offal**
the interior fat of cattle. The viscera and trimmings of a slaughtered animal separated in dressing. Also, the by products of milling used especially for livestock feeds.

- **offspring**
the sons and daughters of parents.

- **oil**
generally a mixture of pure fats which is liquid at room temperature.

- **oil gland**
gland at the bottom of the tail of chickens, ducks,, turkeys (and most wild birds) which secretes an oil used by birds in preening their feathers. Also called *preen gland.*

- **oiling**
the application of oil to the coat of an animal to soften the skin and hair and give the coat a desired lustre.

- **olecranon**
the rounded proximal body prominence of the ulna—the point of the elbow.

- **old process**
pertains to the extraction of oil from seeds.. Same as hydraulic process.

- **oleic acid**
an 18-carbon unsaturated fatty acid (one doable bond) which reacts with glycerol to form olein.

- **olein**
the fat formed in the process of the reaction of oleic acid with glycerol.

- **olfactory**
pertaining to that sense of smell.

■ **olland**
the after effects following the harvesting of hay from clover ley.

■ **omasum**
the third section of the ruminant stomach. Contains a mass of suspended, parallel, rough-surfaced leaves which grind ingesta to a fine consistency. Often called the *many-plies.*

■ **omnivore**
an animal which eats both vegetable and non-vegetable foodstuffs (e.g., the pig), as distinct from a flesh-eating carnivore or a herbivore.

■ **on the hoof**
designating a living meat animal.

■ **onager**
wild ass of Southern Western Asia.

■ **onychorrhexis**
a fracture or fragility of a claw.

■ **onychia**
an inflammation affecting the nails or claws of animals.

■ **oogenesis**
process of development and growth of ova within the ovary.

■ **opaque**
that light can't pass through; neither transparent nor translucent.

■ **open**
a term commonly used of female farm mammals to specify a non-pregnant status. Also, a characteristics of cheeses that have an excessive amount of mechanical openings and/or gas holes.

■ **openness**
a quality in dairy cattle associated with length of body and width and length of rib.

■ **operator**
the person milking the cows by machine.

■ **opetator gene**
a particular area near a structural gene that has the capacity to respond to the regulator gene.

■ **operon**
a collection of genes consisting of an operator gene and the structural genes under its control.

■ **opisthotonos**
a firm of tetanic contraction in which the bead and beefs are bent backward and the body bowed forward.

- **opsonins**

antibodies of blood serum which weaken the cells of micro-organisms so that they are readily ingested, or engulfed, by the phagocytic body cells (white blood cells)

- **optic**

pertaining to the eye or the sense of sight.

- **optimum**

1. ideal;

2. the conditions of temperature, moisture, pH, aeration, fineness etc , which result in maximum growth of a plant or animal.

- **oral**

pertaining to the mouth.

- **orchidectomy**

the surgical removal of one or both testes; castration.

- **orchitis**

an inflammation of a testis.

- **order statistics**

when a sample of variate values are arranged in ascending order of magnitude, these ordered values are known as *order statistics*.

- **orf**

contagious pustular dermatitis. Generally found in older lambs and young sheep in summer, when the casual virus is activated. Characterised by lemons around the mouth and coronet of the hoof, which may be infected by bacteria causing pus formation.

- **organ**

any part of an animal performing a definite function, a group of cells or tissues that are associated in the body to perform one or more functions.

- **organelle**

sub-cellular structures that make up a cell.

- **organic**

pertaining to substances derived from living organisms (referring to carbon-containing compounds)

- **organic acid**

any organic compound that contains a carboxyl group (COOH).

- **organic compound**
a compound of carbon as distinguished from a non-carbon containing compound, designated inorganic.

- **organic evolution**
the sequence of descent of more complex, more advanced, more recent types of plants and animals from the simpler and primitive ones through successive stages in different periods of the earth.

- **organism**
any complete living plant or animal.

- **organogenesis**
the source or growth of the organs of an animal.

- **orifice**
entrance or outlet of a body cavity.

- **orpington**
a general purpose breed of poultry combining both laying and table qualities. Black, buff or white in colour, with white legs and feet, a single comb and hens laying tinted eggs.

- **orthapriea**
inability to breathe except in the upright position. This state is ordinary during dehydration exhaustion.

- **orts**
that portion of an animal's feed which it refuses to eat.

- **osmophilic**
loving a high osmotic pressure. For example, osmophilic yeasts may grow in sweetened, condensed milk.

- **osmosis**
the tendency of two fluids of different strengths which are separated by something porous (a semi-permeable membrane) to diffuse or spread through a membrane or partition until they are mixed. Osmosis is the chief means by which the body absorbs food and is, specifically, the tendency of a fluid of tower concentration to pass through 3 semi permeable membrane into a solution of higher concentration.

- **osmotic equilibrium**
balance in distribution of solutes achieved between two solutions separated by a semi-permeable membrane.

- **osmotic pressure**
  the pressure exerted by the movement of a solvent through a semi-permeable membrane toward equalising solution concentration on oppositc sides of the membrane.

- **ossein**
  a gelatinous protein found in bone.

- **osseous**
  bony.

- **ossification**
  the process of forming bone. Cartilage is, made into bone by the process of ossification. The minerals, calcium and phosphorus are deposited in the cartilage, changing it into bone.

- **osteitis**
  inflammation of a bone.

- **osteoblast**
  a bone forming cell.

- **osteochondritis**
  inflammation of bone and cartilage.

- **osteoclast**
  a cell that breaks down bone.

- **osteofibrosis**
  a form seen mainly in the horse, but it is also found commonly in pigs, goats and dogs. There is a loss of calcium salts from the bones, which become fragile.

- **osteogenesis**
  formation of bone.

- **osteogenic**
  derived from or composed of tissue concerned in the growth or repair of bone.

- **osteoid**
  like bone.

- **osteomalacia**
  a state marked by softening of the bones, muscular pain, loss of weight and weakness. It results from a deficiency of vitamin D or of calcium and phosphorus. An overactive parathyroid gland may also cause osteomalacia.

- **osteophagia**
  literal meaning is 'bone eating' and is a symptom shown by sheep and cattle in certain parts of South Africa which are deficient in phosphorus and sometimes in calcium in soil and herbage.

- **osteoporosis**
a fall in total bone mass. This disorder of bore metabolism occurs in middle life and older age in both men and women. The bone becomes porous and thin due to a failure of the osteoblasts (bone-forming cells) lo lay down bone matrix. This disorder may result from (1) a dietary deficiency of calcium and/or protein, (2) a lowered calcium absorption, or (3) a hormonal disturbance.

- **osteosclerosis**
irregular hardening of bone.

- **otitis**
means inflammation of the ear.

- **otorrhoea**
means a discharge from the ear.

- **out of**
terminology for mothered by in animal breeding.

- **out breeding**
mating animals distinctly unrelated, generally with varied type or production traits. Varies in degree, depending on degree of divergence in type or production traits.

- **outcross**
the introduction of blood from some outside and unrelated source, but of the same breed, into a herd which is more or less related.

- **out-crossing**
system of mating in which several generations of inbreeding are broken by introducing animals not related to the inbred lines.

- **outliers**
livestock kept outside during winter.

- **out wintering**
the practice of keeping cattle in the fields during the winter as distinct from keeping them under cover.

- **ova transplant**
the taking away of an fertilised female reproductive cell from one cow and insertion into another cow for fertilisation and gestation.

- **ovariectomy**
the surgical removal of an ovary.

- **ovariotomy**
  surgical operation for taking away of a diseased ovary.

- **ovary**
  female organ where ova are produced and shed. It is also responsible for the secretion of the hormones oestrogen and progesterone.

- **over-conditioned**
  this is indicated by excess fleshing and irregular fat deposits, especially over the shoulders, back and thighs.

- **over-dominance**
  the interaction of members of a pair of genes (alleles) to produce a phenotypic effect that is superior in the hetero-zygote to either homo-zygote.

- **over-dominance hypothesis**
  the hypothesis proposed by Shull and East in 1908 to explain the cause of hybrid vigour as 'the heterozygosity'.

- **overgrown**
  a trade description applied to wool which has more than normal twelve months' growth.

- **over-milking**
  leaving the unit on the cow after all the milk has been drawn from the udder.

- **over–run**
  the increase in quality of product output over raw material input. In butter, overrun is calculated based on the input of fat and the output of butter. Components of butter other than fat are moisture, silt and curd and these may legally comprise no more than 20 per cent of butter. In frozen deserts, overrun results from the absorption of air which increases the volume of product.

- **over-shot (brachygnathia)**
  a state concerning disparity in the length of the jaws in which the lower jaws appears too short.

- **ovicide**
  any material that kills parasites in the egg stage.

- **oviduct (fallopian tube)**
  fine tubule which transports ova shed at ovulation to the uterus.

- **ovine**
an animal of the sub-family ovidae; goats, sheep.

- **ovine virus abortion**
a contagious disease of sheep characterised by premature birth or abortion near full term, often accompanied with retention of placenta. The disease is caused by a virus of psittacosis-lymphogranuloma-venereum group.

- **oviparous**
producing young from eggs that hatch outside the body.

- **oviposition**
the laying (expelling) of a fully developed egg.

- **ovogenesis**
formation and growth of the ovum.

- **ovoviviparous**
producing eggs within the maternal body. The eggs hatch within or immediately after extrusion from the parents.

- **ovulation**
the detaching of a follicle by the ovary. The ovary of a hen contains a series of follicles (called the follicular size hierarchy). After ovulation, the smaller follicles advance one position in size and re-establish the hierarchy as it existed just before ovulation.

- **ovulation rate**
the number of ova shed into the oviduct at ovulation.

- **ovum**
the female sex cell, produced on the ovary and carrying a simple half of the genes carried by the female in which it was produced. Pl ova.

- **ox ( pl oxen)**
any species of the bovine family of ruminants. Specially, a domesticated and castrated male bovine used for work purpose.

- **oxalacetate**
an transitional compound formed from sugars and sugar-producing amino-adds essential for the oxidation of carbohydrate , fats and proteins for energy.

- **oxidase**
an enzyme that brings about oxidation.

- **oxidation**
chemically, the increase of positive charges on an atom

or the loss of negative charges. There maybe a loss of one electron (univalent 0) or two electrons (divalent 0). The combining of oxygen with another element to form one or more new substance . Burning is one kind of oxidation. Also called oxidation.

- **oxytocin**
  a hormone secreted by the posterior lobe of the pituary gland. It is responsible for activating the let-down mechanism, releasing milk from a cow's udder. It also causes the contraction of smooth muscles in the uterus.

- **oyster shell flour**
  oyster is a hard shell of a shell fish found especially on the bottom of the sea. It contains calcium to the extent of 34-40%, The material is sold either as dried shell or in dried powdered form.

- **packing stations**
  established to collect and market and table poultry.

- **paddock**
  a tiny enclosure or pasture used for grazing and exercising of a single animal like the bull or a mall herd of animals.

- **paddock grazing**
  a scheme of managing grassland in which grazing areas are divided into paddocks by permanent or semi-permanent fences. Each paddock is grazed in turn and then allowed to rest.

- **pail**
  vessel, usually round, of wood or metal for carrying liquids. Milk pail.

- **paint**
  a coat colour of horses (white patches interspersed with darker colours, usually black).

- **paired feeding (food equalising)**
  a method of comparing nutritional effects at an arbitrary low level set by the animal that consumes the least food. Littermates or a set if twins (especially monozygous ones) are considered to be best for paired-feeding studies.

- **palatability**
  the quality characteristics such as flavour, colour and texture of a food product

that make an impression on the organs of smell, sight, touch or taste and have implication in determining the adequacy of the food product to the user.

- **paleontology**
  the study of fossil remains.
- **pallor**
  absence of the typical skin colouration.
- **PABA**
  Para-Aminobenzoic Acid.
- **packer**
  a slaughter and meat-processing businessman.
- **pack house**
  a building in which eggs, etc., are packed into boxes or containers for despatch from a farm to a market, retail outlet, store, etc. Casual workers are usually employed seasonally fur pact house work.
- **palm kernel cake**
  a cake derived from the residue of the kernels of the fruit of the African oil palm (*Elaeis quineesis*) after extraction of the oil. It has a high fibre content and is slightly gritty and is therefore somewhat unpalatable, it is less rich in protein than other cakes and is slow to absorb water. Sometimes used to steam-up dairy cows.
- **palmitic acid**
  a 16-carbon saturated fatty acid.
- **palomitin**
  the fat formed from the reaction of palmitic acid with glycerol.
- **palomino**
  a golden coat colour in horses with the mane and tail a light blond or silvery colour.
- **palpation**
  the means of examining the facade of the body and the internal organs as to their size, position, shape, etc., by the method of feeling with I he band laid upon the skin gently manipulating the structures within reach.
- **palpebral**
  relating to the eye-lids.
- **pan**
  a prefix meaning all or completely.
- **pancreas**
  a gland at the rear of the stomach which produces

insulin and secretes pancreatic juice which passes through the pancreatic duct into the duodenum and there contributes through its ferments to digestion.

- **pancreatic juice**

a thick, transparent liquid secreted by the pancreas into the upper small intestine. It contains the enzymes pancreatic amylase, pancreatic lipase and trypesin; also the hormone insulin.

- **pandemic**

prevalent (as a disease) throughout an entire country or continent or the world.

- **pantothenic acid**

a vitamin of the vitamin B complex that occurs in animal tissues, liver and kidneys. Its absence in the diet or body of rats results in loss of weight and greyness of hair followed by loss of hair. The function and action in man are unknown. It is a pale yellow oil, soluble in water, slightly soluble in ether and insoluble in benzene.

- **panzootiq**

a disease which affects all animals in an area.

- **papanicalaou smear**

a means of discolouration smears of various body secretions from the genitourinary tracts, respiratory , or digestive. It is used lo diagnose cancer or the presence of a malignant process in which exfoliated cells of organs, such as the stomach or uterus, arc obtained, smeared on a glass slide and stained for microscopic examination. It was named for the Greek physician George Papanicolaou who developed it.

- **papiliue**

small nipple-shaped projections located on the interior of the rumen wall.

- **papillomatosis**

a state marked by tumour formation that consists of many nipple-like lesions.

- **papule**

a small, solid elevation of the skin up to 0.5 cm. in diameter. Larger, solid, elevated lesions arc called nodules or tumour. Larger, flat-topped elevations are called plagues.

- **para aminabenzoic acid**

one of the B vitamins. Often

abbreviated as paba.

- **parakeratosis**
any defect of the stratum corneam (horny layer of epidermis) of the skin (especially a condition caused by edema between the cells, which prevents the formation of keratin).

- **paralysis**
in its widest sense may mean loss of nerve control over any of the bodily functions, loss of sensation and loss of the special senses, but the term is usually restricted to mean loss of muscular action due to interference with the nervous system.

- **parameter**
a trait of a statistical populations such as its standard deviation, or its variance. Parameters are seldom measured directly and often can not be but are estimated from samples of the population

- **paraplegia**
a kind of paralysis affecting the movement of and sensation in the legs and the lower part of the body.

- **parasite**
an organism that lives in or on another (the Host) from which it derives food, for all or part of its life. Often, though not always, harmful to the host. Animal parasites include internal worms (e.g., Roundworms, Tapeworm, etc.) and external Flies, Lice, Ticks, etc. Some are parasitic on a secondary host during part of their life cycle (e.g., tapeworm Liver Fluke). Plant parasites include fungi and bacteria (causing such diseases as Mildew, Rust, Fire Blight, etc.), various plants such as dodder and misletoe and numerous insects and their larvae e.g., Aphids, Fruit fly, Eelworms, etc. A semi-parasite is one which feeds partly on its host but also manufactures some of its own food (e.g. mistletoe).

- **parasiticide**
an agent or drug destructive to parasite.

- **parathyroid**
any one of four small glands situated beside the thyroid gland, concerned chiefly

with calcium and phosphorus metabolism.

- **parent stock**
the individual plants or animals from which others have been bred, often in large numbers.

- **parental**
relating to the father.

- **parental combinations**
genotypes and phenotypes like those of the parents in a particular cross.

- **parental generation**
the first generation of a genetic experiment. Individuals in the first generation are often considered to be pure or homozygous,

- **parenteral**
pertaining to administration by injection, not through the digestive (food) tract. I.e., such as subcutaneous, intra-muscular, intra-sternal, intra-venous.

- **paresis**
partial paralysis that affects the ability to move but does not affect the ability to feel.

- **parity**
state of the female with respect to her having produced offspring.

- **parity price**
a guaranteed level of farm prices intended to provide the same approximate purchasing power for farmers as enjoyed during a preceding base period. For example, the-support price for *manufacturing grade milk* might be 90 per cent of the price for tie period 1975 to 1980.

- **parlour**
the house where the cows come exclusively to be milked. As opposed to a cowshed.

- **paronychia**
inflammation or ulceration involving the tissue fold around the claw.

- **parrot mouth**
a malformed mouth of an animal (most common in horses) in which the upper jaw unusually protrudes beyond the lower jaw.

- **parthenogenesis (part biogenesis)**
reproduction by the development of an egg without its being fertilised by a spermatozoon, e.g. drone-bees. It occurs in certain lower

animals and has been observed in turkeys. It docs not occur in mammals,

- **partial correlation coefficient**
the partial correlation coefficient measure the association of the dependent variable $X_1$ with any given independent variable $Y_1$. Thus they are measures of the variation of $X_1$, which is explained by and only by the given $Y_1$.

- **partial dominance**
a situation in which one gene of a pair of alleles is not completely dominant with respect to another. For example, in comprest Hereford cattle an individual that possesses no comprest genes or is of genotype cc is of normal size, those that possess two comprest genes CC are dwarfs and those that possess one comprest gene Cc are midway in between dwarfs and normals.

- **particle**
a minute constituent of matter. The primary particles involved in radioactivity are alpha particles, beta particles alpha particles, beta particles, neutrons and protons.

- **parti-coloured**
breed or variety having feathers of two or more colours, or shades of colour.

- **parturient paresis**
a state, caused by a low blood-calcium concentration which results in partial to complete paralysis soon after parturition,

- **parturition**
the operation of furrowing or giving birth to pigs. It is called calving in cattle.

- **passerine**
having to do with the very large group of perching birds, including more than half of all birds, such as the sparrows, warblers, wrens, chickadees , swallows and thrushes.

- **passive immunity**
the immunity of an animal to disease as a result of receiving blood serum from another animal that is immune to the disease. The immunity thus acquired is not permanent but eventually passes away.

- **passive immunity**
immunity acquired when an individual receives pre-

formed immune products (e.g., antibodies) produced by mother individual.

- **pasta filata**
this pertains to Italian cheeses in which the warmed curd is stretched.

- **pastern**
a part of horse's fool which is located between the fetlock and the coffin joint and corresponds to the first phalanx of the middle finger of man.

- **pasteurisation of swine**
an acute, infectious, febrile, septicaemia disease, characterised by acute fibrinous pneumonia. It is caused by bacteria *Pasterurella multocida*.

- **pasteurisation**
method of killing organisms in a product, usually milk by heating to a controlled temperature which is deadly for the causal agent of a number of milk-transmissible diseases without causing any material change in the natural quality of the substance treated.

- **pasteurise**
method of increasing temperature of milk so as to make it disease free.

- **pasteurised milk**
milk treated to a temperature which kills the majority of the Bacteria present, but without affecting its characters. The treatment applied is that needed to destroy Mycobacterium tuberculosis, the most beat resistant pathogen normally present in cow's milk. Most milk is pasteurised using the High Temperature, Short Time method (H.T.S.T.) by which it is heated to al least 72°C (161°F) for 15 seconds and then rapidly cooled to 10°C (50°F) or less. An alternative is the Holder method by which milk is heated to between 63°-66°C (140°-15°F) for at least 30 min. and then cooled to 10°C (50°F). The adequacy of the treatment and the keeping quality are determined, respectively, by two statutory tests, the phosphatase and methylene blue tests. Pasteurised milk should keep at least 2-3

days. The term is derived from the French chemist, Louis Pasteur, who pioneered the process.

- **pasturage**

all vegetation on which animals graze, including grasses or grass-like plants, forbs and shrubs.

- **pasture**

a fenced part of domesticated forages, generally improved, on which animals are grazed.

- **pasture improvement**

any practice of grazing, seeding, cropping, liming, movin scattering, fertilising, contour furrowing, or other methods of management designed to get better the vegetation for grazing purposes.

- **pasture management**

the application of practices to keep pasture plants growing activity over as long a period as possible so that they will provide palatable feed of high nutritive value; to encourage the growth of desirable grasses and legumes, while crowding out weeds, bushes and inferior grasses.

- **pasture renovation**

the enhancement of a pasture by the partial or complete obliteration of the sod, followed by such liming, fertilising and seeding as may be required to establish more desirable forage plants without an intervening cultivated crisp.

- **pasture supplemental**

a crop used to provide grazing as a supplemental use, as the aftermath of forages cut for hay or seed; or an annual such as Sudan-grass to supplement other pastures. Usually a forage to tide over periods of low pasture production.

- **patch**

an even, circumscribed change in colour of the skin that is larger than 1 cm. in diameter (therefore, a large macule).

- **patella**

the kneecap.

- **patency**

the state of being open or unobstructed.

- **patent**
  unhindered, open , or not closed.

- **paternal**
  pertaining to the father or mate patent.

- **paternal half sibs**
  half-brothers or half-sisters having the same sire but different dams.

- **pathetic nerve**
  the fourth nerve arising from the brain and controls the superior sloping muscle of the eye.

- **pathogen**
  1. a micro-organism, bacterium, or protozoan that: produces disease.
  2. any organism which can cause disease in a particular host or range of hosts. It obtains its nutrients wholly or in part from another living organism,
  3. any agency which can incite disease.

- **pathogenesis**
  the part of the life of a disease producing organism when it is directly associated with a living host.

- **pathogenic**
  that which can produce disease.

- **pathogenomonic**
  a term applied to those symptoms of a particular disease and on which basis the diagnosis depends.

- **pathology**
  the branch of science dealing with disease, particularly with structural and functional changes in tissues and organs of the body affected by disease.

- **pathotype**
  population of pathogen with all the individuals having a particular character of pathogenicity in common.

- **paunch (rumen)**
  the first stomach compartment of a ruminant.

- **peasant farming**
  a scheme of farming in which farmers follow agricultural practices in their own way and are managers and organisers of their farm business. Living and work-

ing are closely related. The entire family of the peasant

has a part in making decisions and executing the farming programme. The peasant farming places a high premium on management and the use of the family labour which maximise farm business income and also allows the organisation of the farm to be adjusted to the capabilities of the various members of the family.

- **peck order**
the social hierarchy in poultry and other birds. The term derives from the habit of dominant individuals pecking submissive ones on the head.

- **pectoral**
pertaining to the breast.

- **ped**
a specifically made light enclosure for packing dead poultry for market.

- **peddock**
a small fenced field used for grazing purposes.

- **pediculicide**
an agent that destroys lice.

- **pediculosis**
invasion with lice.

- **pedigree**
the history of the parents, which should be as complete as possible. A fowl cannot be regarded as pedigree unless it has individual qualities recorded from its parents. May be concerning egg production, showing, etc.

- **pedigree registered animal**
one whose forebears have recorded with a Breed Society.

- **pedigree selection**
the choice of animals in stock-breeding on the basis of desirable characters or qualities in their ancestors.

- **pellet feed**
agglomerated feeds formed by extruding individual ingredients of mixtures by

compacting and forcing feed through die opening by any mechanical process.

- **pellets**
  compressed particles of feed formed by forcing material through die openings.

- **pellicle**
  a thin film, membrane, or skin; a- film on the of a liquid, such as the film that often appears on hot chocolate, which is coagulated whey protein of milk.

- **pelt**
  the natural, whole skin covering, including the hair, or fur of smaller animals, such as sheep and foxes.

- **pelvic bones**
  the two short bones the ends of which are felt close to each side of the vent.

- **pelvis**
  the ring or basin-shaped bone connecting the posterior limbs to the spinal column.

- **pen**
  a small enclosed space for poultry or animals. Also those animals kept in the pen and sufficient in number to fill it. Also called *fold for sheep.*

- **pencilled spike**
  very long and narrow spikes of single comb. A defect that would penalise a bird in the show pen.

- **pencilling**
  small stripes across the feather, either straight or credent-shaped; also refers to *grey marking* of a Rouen duck and black spots on the plumage of Black-red Game hen.

- **pencilly**
  having thin pencil-like spikes on the comb.

- **pendant (pendent)**
  something suspended.

- **pendulous**
  suspended so as to swing freely, such as a loosely attached udder.

- **qualitative traits**
  those traits, such as black and white or polled and horned, in which there is a sharp distinction between phenotypes. Usually, only one or two pairs of genes are involved .

- **quality**
  1. a term for the estimated

description of radiation with regard to its penetrating power. It is usually expressed in term of effective wave length or half life.
2. the term which stands for the excellence of the characteristics as desired by consumers in a product at the time of purchasing.
3. a measure of the degree to which an item confirms the applicable specifications.
4. fineness of tenure; freedom from coarseness in every detail of the body; generally used to indicate superior breeding or merit.

- **quality number**
one of a series of numbers used in English-speaking countries to specify the spinning capacity of wool and determined principally by its fibre fineness.

- **quantitative traits**
those traits, such as skin colour in humans, in which there is no sharp distinction between phenotypes, with a gradual variation from one phenotype to another. Usually, several genes, as well as environmental factors, are. involved.

- **quarantine**
governmental regulations for the prevention of the spread of infectious disease by which an animal or animals, which have come from infected countries or areas, are detained at the frontiers or ports of entrance, or at other official centres, for a period of isolation, before being allowed to mix with stock of the country.

- **quarter**
1. a butcher's term for the limb of an animal and adjacent body parts, detached from a carcase.
2. one of the four divisions of an udder, including one teat,

- **quick freeze**
a means of preserving and storing table birds.

- **quidding or cudding**
the name given to that condition in horses, depending upon injuries to the month or diseases of the teeth, in which food is taken into the mouth, chewed repeatedly and then expelled on the floor of the stall or into the manger.

■ **quill**
the hard stiff feathers of the wings and tail.

■ **quill bond**
preservation of the web of a feather in its sheath.

■ **quintuplets**
five offspring produced at the same birth or one of five who have been born at one birth.

■ **quotidian**
recurring every day, such as malaria fever.

■ **r**
abbreviation for repeatability of a sire summary or a cow index.

■ **rabbit**
a device used to move a sample faster from one place to another, e.g. in a research reactor.

■ **race**
1. a while streak on an animal's face.
2. a term used by breeders for the descendant of a common ancestor which exhibit common characteristics, but which differ slightly from other individuals or races within the species, also used generally to mean a breed, stock, group, class or genus of plant or animal,
3. a passage along which livestock may be moved (e.g. from pen to dipping tank), normally made of gates set in two parallel lines sufficiently apart to lake one animal at a time.

■ **rack**
the gait of a horse in which only one foot touches the ground at any one time, producing a four-beat gait. The legs move in lateral pairs but not quite in unison, so that that foot is lifted and put down alone.

■ **rack up**
to fit a rack with hay or straw.

■ **radiant heat**
heat transmitted by radiation (like that of the sun) as contrasted with that transmitted by conduction or convection.

■ **radiation**
the way of emitting radiant energy (heat) in the form of waves or particles. The solar energy is transferred by radiation. It is also an important method of heat loss from an animal to cooler

objects and heat gain by the animal from warmer objects.

- **radiation sterilisation**
the method in which radiation is used to cause a plant or animal to become sterile or incapable of reproduction. Also, the use of radiation to kill all forms of life (specifically bacteria) in food on equipment, etc.

- **rooster (cock)**
an adult male chicken.

- **ruot**
mean square deviation. The square root of the second moment of a set of observations taken about some arbitrary origin.

- **rop**
the abbreviation of record of performance, a breeding stage in the national poultry improvement plan.

- **ropey wool**
badly scoured wool which has become slightly jangled, felted and knotted in processing.

- **rose comb**
a good broad comb, nearly level across the top and covered with small points.

- **rotation pasture**
a meadow crop grown as a part of a planned cropping system.

- **rotational grazing**
a system to manage grassland in which successive areas are intensively grazed for a continuous time period, in which defoliation is rapid and complete, followed by a rest period to provide re-growth time. This system is followed in a lot of farms to match the requirement of the livestock with grass growth.

- **rotor**
  the internal turning component of a vacuum pump which holds the vanes in machined slots.

- **rough grazing**
  a term generally applied to grassland of poor quality with a low stock-carrying capacity, particularly that covering rough land, heath, moor land and hill areas, where cultivation and the use of fertilisers is impractical and which is characterised by coarse grasses, sedges, heather, bracken or mosses, etc. such hind is usually grazed by hardy bill breeds of sheep and some cattle.

- **rough stalked meadow grass**
  a type of close-growing perpetual grass (*poa trivialis*) with short creeping stolons, a rough, flat stem and an erect panicled flower head, reddish, purplish or green in colour. It is common in lowland meadows and pastures, particularly on rich moist soils. it is included in seeds mixtures for permanent pastures, particularly those on wet heavy soils and grows vigorously producing a low closely knit sward which remains green through the winder. it is very palatable to livestock and i also useful for hay,

- **roving**
  the last phase of the drawing process, resulting in a strand of wool of the desired thickness for spinning into yarn; also the name for the strand itself.

- **rrhaphy**
  this suffix refers to an operation in which some opening or tear is closed by stitches.

- **rubarth's disease**
  acute contagious disease affecting mainly young dogs, characterised by fever, vomiting, diarrhoea and sometimes convulsions. It is caused due to filterable virus.

- **rubbers**
  a terra for sheep scab.

- **rugged**
  refers to a bulky physically powerful animal.

■ **rumen**
the biggest of the four stomach sections in the adult ruminant and the one to which most cellulose goes first. The spot of active bacterial digestion. It contains a quantity of feed and water equal to approximately one-seventh of the ruminant's body weight. It is also called paunch.

■ **rumen flora**
the micro-organisms of the rumen.

■ **ruminant**
one of the order of animals that has a stomach with four complete cavities, rumen, reticulum, omasum, abomasum through which food passes in digestion. The animals chew their cud; they include cattle, sheep, goals, deer, antelopes, elk and camels.

■ **ruminate**
to regurgitate previously eaten feed for further chewing. To chew a cud.

■ **ruminating**
chewing the cud; remastication of a bolus of feed which has been regurgitated from the rumen of the animal.

■ **rumination**
the casting up of food (cud) to be chewed a second time, as in cattle. a chewing of the cud, as by ruminants.

■ **run**
an enclosed space for animals, mainly poultry, in which they arc free to roam and feed.

■ **rye bran**
the crude outer covering of the rye kernel as removed in the usual processes, other than scouring.

- **rye middlings**
  a small proportion of fine bran particles, germ and a large proportion of low grade fibrous flour as separated in the usual process of flour milling. the crude fibre content docs not usually exceed 5 per cent.

- **rysamin B**
  a proprietary syrup made from rice-polishing, in vitamin B therapy.

- **sacral**
  pertaining to the sacrum.

- **sactocoma**
  tumour that develops from tissues of mesoderm origin.

- **sacrum**
  the fused mass of vertebrae found in the hip region, the fourth region of the spinal column.

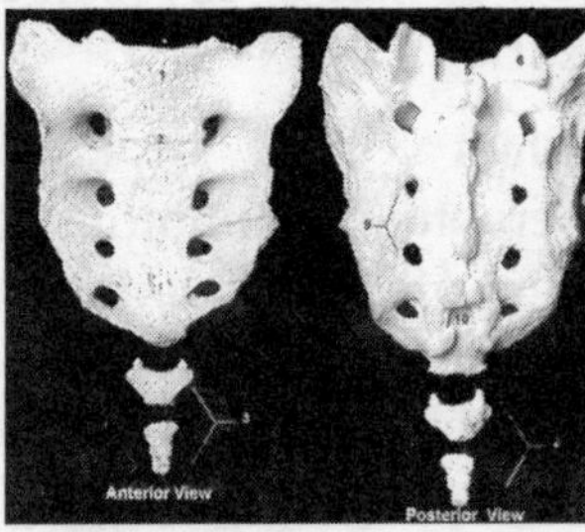


- **saddle hackle**
  the pointed feathers banging from the saddle over each side of the back.

- **sagittal**
  the term applied to a structure or section running transversely across the trunk or a limb.

- **sagittal plane**
  interior posterior plane or segment paralleled to the long axis of the body.

- **saflower meal without hulls**
  remains of saflower seed after most of the hull and the oil has been removed. Protein content is more than 40% but the quality is not as

good as that of soybean sail meal.

▪ **sainfoin**

a perpetual leguminous fodder plant (*onobrychis vicifolia*)) once widely grown for both hay and silage, with aftermath grazing, mainly in chalky areas, usually on its own, but sometimes included in seed mixtures. it has a similar nutrition value as lucerne which it is sometimes substituted for on potash-deficient soils, being somewhat drought-resistant. It has a strong, woody, long foot, a branching stem with pinnate leaves, pink flowers in a conical raceme and single-seeded pods.

▪ **saliva**

a dear, fairly viscid solution secreted by the salivary glands into the mouth. It contains the enzymes salivary amylase and salivary maltose

▪ **salivary**

of or pertaining to saliva

▪ **salivation**

an excessive secretion of saliva.

▪ **salmonella**

a genus of bacteria. Many types exist some of which cause infections diseases (*salmonellosis*) in livestock. Important types are *s. dublin* and *S. typhimurium* which often affect cattle, causing a variety of symptoms including diarrhoea (at times bloody), fever, loss of appetite, lower milk yield, abortion, sometimes pneumonia and death, etc. infection tan be transmitted in faeces, in slurry spread on grassland, in contaminated milk or animal feeds, by healthier carriers, cross-infection at markets, etc., *S. typbimurium* and certain other types arc the cause of food poisoning in man.

▪ **season**

1. the oestrus or 'heat period of a female animal.
2. one of the four climatic divisions of the year, e.g. spring.

▪ **seasonal average**

an arithmetic mean of the monthly or quarterly values over a period of years.

- **seasonal incentive plans**
payment methods devised under certain federal orders to encourage more even production of milk throughout the year.

- **silage crop**
crops which are preserved in a moist form by partial fermentation in a tight receptacle, e.g. jowar, etc.

- **sinistral**
of or pertaining to the left side: left; or left-handed.

- **sinus**
a cavity in a bone of skull which communicates with the nostrils.

- **sire**
the male parent. To father or to beget.

- **sire indices**
various methods of calculating the capabilities of bulls to transmit economically important production traits.

- **sire summary**
an approximation of the genetic transmitting ability of a bull.

- **spiv ewe**
an ewe in poor state and which will not fatten.

- **stained pieces**
a trade term which refers to urine-stained wool and small portions of yolk-stained wool from the clutch.

- **stained wool**
wool discoloured so that it cannot be made white by scouring. It can only be used in dark-coloured fabrics. Nearly all fleeces contain some wool that has been stained by urine. In the process of skirting or sorting this naturally stained wool is removed from the fleece and put into a soil by itself, called stained pieces. It has a dark brown, scorched appearance after scouring.

- **stale**
a stage during which an animal does not work or lactate at the normal standards, as opposed to loom. also refers to milk products that have deteriorated in storage.

- **stall**
an enclosed compartment in a cowshed or stable in which cattle or horses stand or lie down.

- **standard bred**
  birds bred to conform to the official standards for a breed or variety.

- **standard curve**
  a line constructed between points that represent values determined in analysing a reference substance. For example, id the dye-binding test for milk protein, values are often based on a standard curve constructed from data obtained in performing kjeldahl tests on milks of varying pro-t' in content,

- **standard error**
  an estimation of the standard divergence of means of samples drawn from some single (statistical) population. Often calculated from a single set of samples.

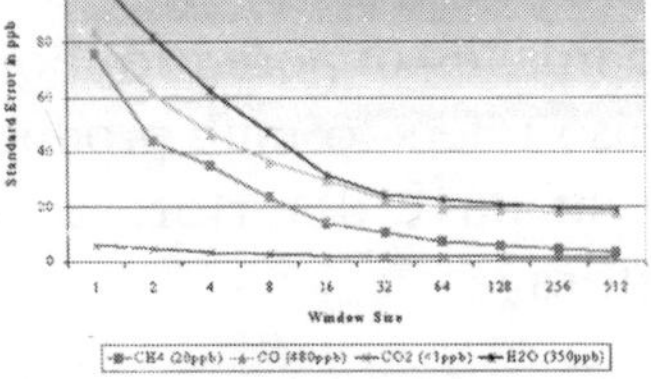


- **standard of perfection**
  set of stipulations for exhibition poultry.

- **standby pool**
  an deal between cooperatives in markets that have periods of milk shortages and cooperatives in areas of surplus milk production. Cooperatives in the deficit market make monthly payment is to certain cooperatives in the surplus region (particularly Wisconsin and Minnesota, U.S.A) to assure that they have sources of milk available at reasonable prices when local milk does not supply their class 1 needs.

- **standings**
  the raided parts of the floor of a cow house on which the cows are kept in pairs in stalls, each normally about 1.5 m(1ft.) deep and 2.1 m (7 ft.) wide.

- **staple**
  wool fibres naturally binding together in a fleece forming a lock. the staple length of wool is n indication of the average length of the wool fibres.

- **staple length**
  the length of wool obtained by measuring in the natural staple without stretching

the crimp out of that individual fibres to obtain the fibre length.

- **staple wool**
  a fine, strictly combing wool from the western half of the United States. It corresponds to the delaine wool from the central part of the united states,

- **star lots**
  packages or lots of wool consisting in New South Wale and Queensland of four bales and under and in Victoria, South Australia and Tasmania of three bales and under, listed in a different catalogue and sold in a different sale room.

- **starch**
  a polysaccharide having the formula ($C_6 H_{10} O_5$)n. an important source of energy for livestock. yields glucose upon complete hydrolysis.

- **starch equivalent (SE)**
  a net energy system of feed assessment used widely in Germany and other European, countries. one kilogram of SE is equivalent to 2356 kcal net energy for fattening (NE*f*). The system is based on work by kellner and his successors. SE values are calculated based on digestible nutrients and crude fibre of the diet.

- **starvation**
  the withdrawal of an animal of any or all the food elements essential to its nutrition

- **stasis**
  a term applied to stoppage of the row of blood in the vessel or of the food materials in the intestinal canal.

- **slate milk order**
  state government regulated farm, whole-sale and retail milk price.

- **station**
  a large sheep ranch (in Australia)

- **statistically significant**
  having value in a statistical context. This generally specifies in research to tests for distinctions which are the outcome of a treatments. The reliability of such differences is expressed as degree of probability or the percentage of time an observation from a normal population would he expected lo tall outside a certain range of variation from the mean.

- **statistics**
  mathematical operation by which a given quantity of masses is reduced to a meaningful data which can be readily interpreted in order to draw conclusions concerning the original data.

- **stearic acid**
  an 18-carbon saturated fatty acid which reacts with glycerol to form stearin.

- **stearin**
  the fat formed from the reaction of stearic acid with glycerol.

- **steely**
  a term applied to wool lacking quality and possessing a characteristic 'steely' or 'glassy' sheen. It is produced on pastures known to be deficient in certain trace elements.

- **steer**
  a male animal among cattle that was castrated before the appearance of the male secondary sex characteristic.

- **stell**
  an open stone or uneven metal enclosed space or shelter for sheep and cattle, usually circular, built hilts and moor land.

- **stopping**
  the tops of the secondary feathers in bird's wing, which are shorter and shorter as they draw near the back, giving the idea of steppes.

- **sterile**
  free from living micro-organisms. also, incapable of producing young.

- **sterilisation**
  the process of freeing a substance or an article from all living organisms.

- **sterilise**
  to take away or kill all living organisms. Also, to make barren or unproductive, as a vasectomy in bulls.

- **sterilised milk**
  this is a special type of liquid milk sold for public consumption. This milk must pass the turbidity test after prolonged, heating, sometimes for up to 30 minutes, to between 110° and 115°c (230°-240°f). This heating, carried out in sealed airtight containers in which the milk is later sold, kills all the bacteria present and further contamination is prevented by the airtight

conditions. Such milk has a stored life of at least 7 days but the sterilising process affects its flavour and colour.

- **sterility**
  not capable of reproduction.
- **sternum**
  the breastbone.
- **steroid hormones**
  hormones of relatively simple molecular arrangement such as progesterone, oestrogen and testosterone.
- **sterol**
  one of a class of complex, fat-like substances widely distributed in nature,
- **stibbler**
  a horse set free to graze on stubble.
- **stibestrol**
  drug used in cattle feeding which stimulates gain and iced efficiency with estrogenic-like activity.
- **sticking**
  a term used at abattoirs for weeding livestock, by culling their threats, after they have been stunned and hoisted.
- **stifle**
  the knee joint of a horse.
- **stifle joint**
  the joint between tibia (shinbone) and thighbone in the bind leg of a horse.
- **stilbestrol**
  it's a kind of drug used in cattle feeding. This drug stimulates gain and feed efficiency with estrogenic like activity.
- **stillborn**
  1. an offspring which has born lifeless;
  2. dead at birth.
- **stimulant crops**
  these are crops whose utilisable part stimulate the nervous system, e.g., tea, coffee, tobacco etc.
- **stock sires**
  selected male birds for breeding.
- **stock vaccine**
  vaccine prepared with germs obtained from animals other than the ones to be immunised or from cultures.
- **stocking**
  distinct coloured feather or hair on an animal's leg.

- **stocking rate**
  the number of grazing animals that arc allowed to graze a given area of pasture, normally expressed in terms of livestock units per acre or hectare.
- **stockman**
  a man in charge of livestock.
- **stomach**
  that part of the digestive tract lying between the oesophagus and the small intestine, a 4-compartment organ in ruminants; a single compartment organ in non-ruminants.
- **stomatitis**
  this refers to an inflammation of the mucous membrane of the mouth.
- **stool**
  means faecal material or evacuation from the digestive tract.
- **stop cock**
  a screw down valve which will hold against pressure.
- **storage drying**
  its a special means for drying bales of hay in a barn over a ventilated floor using a how of unheated air. Initially about four layers ate dried and subsequently further layers are added and dried until the barn is filled. The bales are then stored *in situ*, also called deep storage drying and barn hay drying.
- **store**
  an animal kept at a steady rate of growth prior to later fattening for market.
- **store animals**
  animals fed at a nutritional level which allows only very slow growth, usually in winter.
- **store condition**
  refers to a sheep which is in good muster; a sheep that is not the property of the holding on which it is found.
- **stover**
  it's a kind of mature, cured stalks of such crops as corn or sorghum, from which the grain has been taken away. It can also be called a roughage.
- **straw yard**
  an enclosure covered with straw as litter, in which cattle are kept during winter.
- **streptococcus cremoris**
  this is a species of gram-

positive cocci that ferments lactose. They divide in one plane only like rods. They arc used extensively in manufacturing cultured dairy products and cheeses. *Streptococcus lactis* is closely related species for the same purpose.

- **stress**
the sum of all non-specific biological phenomena caused by adverse conditions or influences. It includes physical, chemical and/or emotional factors to which an individual fails to make satisfactory adaptation and which cause physiological tensions that may contribute to disease.

- **stride**
the distance from one footprint of a horse to the prim of the same foot when it next comes fully to the ground.

- **strike**
it's a kind of infestation of the flesh of sheep by maggots hatched from the eggs of the blow fly laid in the fleece. The condition, in which sheep arc said to have been 'struck', causes intense irritation and death can occur within a week. Protection is afforded by dipping.

- **stringy wool**
thin small-bodied staples, which hang loosely, are generous in condition, show a clear crimp and are usually perished at the tip.

- **strip**
the last milk obtained from the milking process which is high in butterfat.

- **strip cup**
a small container into which the first few squirts of milk are drawn by hand from a cow's teats before milking starts. This foremilk is discarded as it has a high bacterial content.

- **stripping**
the last drops of milk drawn from a cow's udder during milking.

- **strobilation**
this is a normal kind of asexual reproduction in which segments of the body separate to form new individuals, as in tapeworms.

■ **stroma**
it's a connective tissue comprising the framework of an organ.

■ **strong**
a term referring in wool which has comparatively bread fibre diameter for us breed or type.

■ **stud**
a unit of selected animals kepi for breeding purposes, as of bulls and horses. Abbreviation for stud hore; a stallion,

■ **stud mating**
practice of placing females at intervals in a special mating pea in which the male is confined.

■ **sturdy**
this is a unique disease of sheep and sometimes cattle, characterised by faintness or stunning, caused by tapeworm cysts in the brain membranes. The cysts are formed by embryos of the dog tapeworm, *Taenia caenurus*, hatched from eggs deposited in grass in the faeces of infected dogs and subsequently eaten during grazing.

■ **sty**
a pen in which swine are fed and housed.

■ **subclinical**
a disease condition without clinical manifestations,

■ **subcutaneous**
situated or occurring beneath the skin.

■ **subcutaneous connective tissue**
refers to the skin laver beneath the dermic.

■ **sub-humid climate**
climate with adequate precipitation to support a moderate to dense growth of tall and/or short grasses usually found in areas with 50-75 cm. of rainfall.

■ **succulence**
a condition of plants characterised by juiciness, freshness and tenderness, making them appetising to animals.

■ **succulent**
soft, watery feeds relatively low in fibre.

■ **succulent foods**
feeding-stuffs with a relatively high water content, approximately 70-90%. They are highly digestive but are low in oil, protein and fibre. They are valued particularly for winter feeding to balance dry fodders which tend to cause constipation and especially for pregnant, sick or high yielding animals. Examples include the carbohydrate-containing tubers and roof crops such as potatoes, carrots, mangels, swedes and turnips etc. various by-products with a low dietetic value are sometimes regarded as succulent foods, such as wet brewers' grainy wet sugar beet pulp, liquid skim milk, etc.

■ **sucker lamb**
a lamb that is sucking its mother and sufficiently heavy for slaughtering.

■ **sucker cow**
a cow which is allowed to rear its own calf and is then used for beef production, as distinct from being used as a dairy cow for milk production.

■ **sucking pig**
a youthful milk-fed piglet.

- **suckle**
  to nurse at the breast or mammary glands.
- **surcease**
  an enzyme present in intestinal juice which acts on sucrose to produce glucose and fructose,
- **sucrose**
  a disaccharide having the chemical formula $C_{12}H_{22}O_{11}$. It hydrolyses to glucose and fructose. Generally known as *cane, beet, or table sugar*.
- **sudoriferous glands**
  glands in the skin which are closely associated with primary follicles and secrete the suint. It is usually found in sheep.
- **suint**
  the dried up perspiration found in the natural fleece. It is easily soluble in water.
- **sulphur (s)**
  a non-metallic chemical element, occurring as pale-yellow crystals in numerous forms, required by living oıganisms a constituent of protein (particularly keratin in animals) and some oils.
- **summer feeding**
  the feeding system of cattle in the summer months involving outdoor grazing with or without supplementary1 rations.
- **sun burn**
  some cattle with a white coat colour and pink skin, which have been bred in temperate climates, show a diffused superficial inflammation of the skin when imported into hot countries.
- **sun cured**
  dried by exposure to the sun.
- **super**
  a prefix indicative of 'above' or 'excess'.
- **superfetalion**
  two pregnancies at different times in the same conception period of the female.
- **superficial**
  pertaining to or situated near the surface.
- **superior**
  uppermost in animal anatomy.
- **super ovulation**
  detaching of an unusually large number of ova.

■ **supplement**
extra feed used to improve a common ratio such us pasture.

■ **supplemental pasture**
generally, a crop planted mainly for another purpose but used to provide grazing as a supplemental use.

■ **supplementary rations**
the concentrates red to livestock in addition to bulk, foods such as hay straw and roots.

■ **supplements**
technical products for use at less than 5 per cent of the total ration, in which they are included and designed to supply planned proportions of vitamins, trace minerals, one or more non-nutrient additives and other special ingredients.

■ **suppository**
an easily unfeasible medication for the urinary, reproductive, or lower intestinal tracts.

■ **suprise tests**
tests performed for verification on cows whose projected 2 x (twice daily milking), 305D(305-day lactation). ME (mature equivalency) records exceed certain levels of milk and/or fat as established by breed associations. Suprise tests include the entire herd. When possible, suprise tests are made by a supervisor other than the one who normally tests the herd.

■ **suramin**
a drug used against trypanosomes.

■ **surface colour**
the part of the feathers exposed to sight sometimes called top colour.

■ **surplus milk**
the amount of grade. A fluid milk in excess of that needed for class i purposes.

■ **surrey chicken**
term used to explain a table bird especially fed and fattened and white fleshed.

■ **vulnerability**
inverse of resistance. Vulnerability.

■ **suspension**
a pharmaceutical preparation in which ingredients are dispersed as visible particles and which is consequently turbid.

**suspensoid**
an apparent solution that is seen by the microscope, lo consist of small particles of solid dispersed material in active Brownian movement.

**suture**
to stitch; or series of stitches to secure the edges of a wound.

**swampy back**
a depression in the back between the shoulders and the loin,

**sward**
to many, sward, turf and sod are nearly synonymous. sward usually refers to the grassy surface of a pasture, lawn or playing field; turf and sod to the stratum of earth filled with routs as well as the surface.

**swarm**
the simultaneous emergence or assembly in one location of large numbers of insects, often 10 establish a new

**swatch**
a collection of grass, corn or other crop cut by a mower or scythe, lying on the ground.

**swayback**
most common in lambs at birth or a few weeks old, caused by progressive destruction of the while mailer of the lamb's braid. while it is in the womb of an ewe with low blood copper content. Mild cases may survive and copper therapy is sometimes effective.

**sweet butter**
not salted butter.

**sweetened condensed milk**
the food obtained by partially removing water from a mixture of milk and a safe and suitable nutritive sweetener.

**swine erysipelas**
an infectious bacterial disease infecting pigs, birds and fish. it is cased by *Erysipelothrix rhosicpoihiae* which is non-motile, non-capsulated, gram positive bacteria measuring 0.5 to 2-5u in length and 0.3 to 0.6; in width.

**swine fewer**
an infectious noticeable disease of pigs caused by a virus characterised by refusal to eat, fever, discharges from the eyes, foul-smelling diar-

rhoea, distressed breathing and general weakness. the disease may be acute, particularly to young pigs and death may occur in a few days or it may take a chronic form, mainly in older pigs, which remain ill for a long period and lose condition, but may not die. Also called *hog cholera* and *pig typhoid*.

- **swine paratyphoid**
  infectious bacterial disease occurring in acute op chronic form characterised by fever, anorexia and red or purple discolouration of skin on certain parts of the body fallowed by diarrhoea. it is caused by *salmonella choleraesuts* which is specially adapted to swine.
- **swine vehicular exanthema**
  an acute infectious disease of swine characterised by the formation of vesicles on the snout, mouth cavity, feet and sometimes on other parts of the body. It is caused by a filterable virus of 13 to 20 $m_i$ size
- **swirl**
  hair on an animal that grows in opposite directions (swirl) to the rest of the hair on the body.
- **switch**
  the brush of hair on the cod of a bovine's tail
- **symbiosis**
  the living together in intimate association of two dissimilar organisms, with a resulting mutual benefit.
- **symmetry**
  a balanced development of all parts,
- **sympathetic nervous system**
  part of the automatic nervous system which, in general, has the opposite effects of the other pan of the autonomic nervous system, the parasympathetic nervous system,
- **symphysis**
  a line of junction— an immovable joint between two cartilage bones.
- **symptom**
  the manifestation of disease symptoms maybe observed by an outsider, or subjective which could be perceived by the patient only.
- **syn**
  a prefix indicative of union.

- **synapsis**
the pairing of a homologous set of chromosomes during the first meiotic division, the chromosomes occur as paired chromatids joined at the centromere.

- **syndactylism**
webbed toed form in the chicken.

- **syndicate**
1. a group of breeders who have common (usually financial) interest in an animal frequently for purposes of getting the animal progeny tested.
2. a combination of signs and symptoms which. characterise a particular disease or abnormality.

- **synergisis**
a substance that increase the effect of another substance, although its effect is negligible when used alone. Lecithin, citric acid and phosphoric acid are examples of substances that act synergistically to prolong the keeping lime of fats containing natural or added anti-toxidants. Sesames is a sinergist for pyrethrum in insecticides.

- **synergism**
1. the ability of two or more organisms to bring about a change (usually chemical and biochemical) that neither can accomplish alone.
2. the term is also used to indicate the effect of two or more antibiotics acting independently.

- **syngamy**
fusion of identical gametes.

- **synsotosis**
the term applied to a union by bony material oil adjacent bones usually separate. it may occur in the spinal column in old animals.

- **synotia**
the (virtual) absence of head in a stillborn animal.

- **synovia(synovial fluid)**
a viscid fluid confining synovin, or muscin and a small proportion of mineral salts. it is secreted by the synovial membrane and resembles the white of an egg. It is contained in joint cavities, bursae and tendon sheaths.

- **synthesis**
the bringing together of two or more substances to form a new material.

- **synthesising**
  utilising various processes and/or operations necessary to build up a compound. also, it is processing to build complex compounds from elements or simple compounds.

- **syphilis**
  an infectious disease which spreads mainly through sexual intercourse.

- **syringe**
  a hydraulic device used for drawing up and measuring fluids and administering them mostly by injection.

- **syrup feeds**
  by-products of the whisky distilling industry which are turned into animal feeds-particularly for dairy cattle. they comprise a blend of malt or maize draff and distiller's syrups and have been shown to contain high sources of energy and protein.

- **system of farming**
  the combination of products on a given farm and the methods or practices that are used in the production of these products is known as system of farming.

- **system of mating**
  process which causes the non-random combining of genotypes to change the genotypic frequencies in the next generation but not the gene frequencies; common are crossbreeding arid various forms of inbreeding.

- **systematic error**
  an error which, for known changes in field conditions, undergoes proportional changes in magnitude and which, for unchanging conditions, remains unchanged, both m sign and magnitude. because for given conditions the sign of the systematic error remains the same, the separate errors in a series of measurements will be cumulative in their total effect.

- **systemic of**
  pertaining to, or involving the body considered as a functional whole, as la systematic circulation.

- **$T_2$ toxin**
  poison which may kill cattle or poultry after eating stored corn containing the fundus *Fusarium tricinctum*.

▪ **table birds**
poultry raised for their meat us distinct from for egg production.

▪ **table birds (broiler)**
resulting from crosses between White Cornish Game (derived from Indian game) males and white Plymouth Rock females.

▪ **tachycardia**
a disturbance of the heart's action which produces great acceleration of the pulse (pulse rate above 100 per minute in man).

▪ **tachypnoea**
an increase it the rate of breathing due to some pathological condition.

▪ **tack**
1. pasture rented for grazing livestock.
2. riding equipment, such as the bridle and saddle, also refer to equipment used in the fitting and showing of animals,

▪ **tactile**
pertaining to the sense of touch.

▪ **tag**
a metal disc or strip with letters, figures, or characters stamped on it, placed in the ear of an animal for individual identification.

▪ **tags**
heavy, dung-covered locks, usually called *dags* in great Britain.

▪ **tail coverts**
soft curved feathers at the sides of the lower part of the tail of the male bird.

▪ **tail feathers**
the stiff and straight feathers of the tail; sometimes the two top ones are slightly curved, but they are generally straight and in a cockerel are contained inside the sickles and tail coverts.

▪ **take**
to accept a male in coitus. to result in a mild infection after vaccination.

▪ **tallow**
the fat extracted from adipose tissue of cattle and sheep.

▪ **tanbark trail**
a term commonly associated with those who exhibit animals in competition at fair and shows. Tanbark is the bark of several trees (oak, chestnut etc.). used as

a ground covering in citrus lots, race tracks, livestock pavilions, etc.

- **tandem method of selection**
  exertion of selection pressure first on one trait, then on a second and so on, rather than selection for all traits simultaneously.

- **tankage**
  a protein supplement used as an animal feed- it consists of ground meat and bone by-products of animals that have been slaughtered.

- **tanker load**
  in milk collection and distribution, large quantities of milk in a large stainless steel tank borne on a transport truck— the method of transporting milk.

- **tannin**
  a substance that has an astringent, bitter taste.

- **tapeworms**
  a class of parasitic flatworms which inhabit the small intestine of animals and are named from their dirty white tape-like appearance. They consists of a ribbon-like chain of segments anchored at one end and trailing through the gut, reaching several feet in length in some cases. They seldom require treatment in livestock.

- **tapioca chips**
  thin sliced pieces of tapioca tubers in dried condition.

- **tapioca flour**
  tapioca has different names in different parts of the world. The English speaking territories refer to it as cassava or cassava root- in Latin American it is known a *maniocs*. The dried tapioca root meal contains about 1% protein, 1.3% fat, 85% nitrogen- free extract and 1.8% fibre.

- **tapioca spent pulp**
  consists everything except the starchy portion of the tapioca tubers which has been removed.

- **tapping**
  the popular name for the withdrawal of fluid from body cavities, or the subcutaneous tissues of the body.

- **tare**
  the weight of the bag or bale-pack which contains wool. it is almost universally

the custom to deduct the tare so that the buyer only pays for the net weight of the wool.

- **tarsal**
  pertaining to the tarsus.

- **tarus**
  the joint in the hind legs of certain quadrupeds, e.g., in horse.

- **taste**
  one of the special senses and involves certain papillae found mainly on the tongue. In the crevices of the papillae there are certain small organs which are called taste buds and it in these buds that the sensations are first received. Each of these buds possesses a number of minute projections from its free surface. These are the ending of nerve-fibres which receive impressions from dissolved food or other material which is presence in the mouth and convey them to the brain.

- **tattoo**
  permanent identification of animals produced by placing permanent ink under the skin; generally put in the ears of young animals.

- **taxon**
  a general term for any taxonomic rank, from sub-specific to divisional. pl *taxa*.

- **taxonomy**
  the science of biological classification of organisms, based as far as possible on natural relationship like mutual affinities or similarities. Taxonomy provides a logical informative system of naming and a key for the identification and classification of unidentified organisms.

- **teart**
  molybdenosis of farm animals caused by feeding on vegetation grown on soil that contains high levels of molybdenum.

- **teart land**
  the rich pasture of Somerest on heavy soils containing an excess of molybdenum renowned for causing scour in sheep and cattle grazing on them.

- **tease**
  to stimulate an animal to accept coitus.

- **teat**
the projecting part of the udder.

- **teat cup**
one of the four tube-like devices forming the cluster of a milking machine. Each test cup consists of an outer metallic shell lined with synthetic rubber, which firs over the teat and into which milk is drawn by the application of a partial vacuum.

- **teat dipping**
a process of applying disinfectant to a cow's teats as a precaution against the spread of mastitis, involving the dipping of each teat into a cup containing the disinfectant. The three conventional designs of cup are (a) a straight cup, (b) a cup on lop of a 'squeeze bottle and (c) a non-squeeze, two chamber model. The most effective cups are at least 1.5 cm. (3 in.) deep, which ensure complete immersion. Antiseptic may also be applied by sprayers but this is regarded as less-effective. The three main disinfectants used are iodophors, hypo chlorite and chlorhexidenes and they are applied after the udder, but prior to milking.

- **teat meatus**
small canal located in the end of each teat; also called streak canal. The teat meatus is approximately one centimetre long and is kept closed by sphincter muscles. It prevents the escape of milk between milking and serves as a barrier to the entrance of micro-organisms and foreign material; its size influences rate of milk withdrawal (fast-milking cows usually have a teat means of larger diameter than do slow- milking ones).

- **ted**
to toss swaths of new-mown grass in haymaking in order lo expose more grass to the sun and air and quicken the drying

- **tedder**
a tractor-drawn haymaking machine, usually used soon after a grass crop has been cut to loosen and aerate the swath.

- **teg**
  a sheep of two years of age.

- **telegony**
  disbelief that previous dealings affect future offspring, e.g., cows bearing calves by one sire will have future calves influenced by him even though future calves arc by different sire,

- **teleology**
  the theory that all organisms or their parts have evolved for a particular purpose.

- **telophase**
  the fourth stage of mitosis characterised by an elongation of the chromosomes, the disappearance of thespindle fibres and the reorganisation of the nuclear membrane.

- **temperament**
  disposition of an animal.

- **temporary environmental effects**
  environmental effects which impinge on an animal for relatively short periods such as a sudden weather change or a minor injury.

- **temporary grassland**
  arable land sown to a leys for a limited period of years, as distinct from permanent grassland.

- **temporary pasture**
  vegetation grazed for a short period only; not more than one full year

- **tend**
  to care for, as a flock of sheep.

- **tender**
  a term applied to wool that will break relatively easily when tension is applied.

- **tendon**
  the strong tissue terminating a muscle and attached to a bone, for leverage purposes.

- **term**
  the gestation period.

- **terminal disinfections**
  disinfections of a room or building after it has been vacated by a patient who had a contagious disease.

- **terminal infection**
  an infection with pathogenic organisms occurring during the course of a disease that terminates in death.

▪ **territory**
the part of the home range of an animal which it will defend against other members of its own species (so that maximum utilisation of an area's amenities is achieved by the species).

▪ **territory wool**
those wools chat grow in the slates west of the Missouri river. wool from Texas, California and Oregon are frequently placed outside the territory group, as are also the small amounts produced in Eastern Nebraska and all of Kansas and Oklahoma.

▪ **teschen disease**
a contagious viral disease of pigs, characterised by generalised muscular tremous, hyperaesthesia, convulsions frequently followed by paralysis of extremities, coma and finally death caused by a filterable virus with the particle size of about 28 m.

▪ **test cross**
a mating test to determine if an individual is a carrier of a recessive gene, homologous chromosomes . Each of which consists of a pair of identical chromatids linked at the centromere,

▪ **test of normality**
a test of a set of observations to see whether they could have arisen by random sampling from a normal population.

▪ **test of significance**
a method of making due allowance for the sampling fluctuation affecting the results of experiments or observation.

▪ **tester**
one who tests milk.

▪ **testicle**
male sex organ which produces sperm alter sexual maturity. Pl. *testicles*.

▪ **testis**
one of the paired reproductive glands in the male which produces spermatozoa and secrets male sex hormones.

▪ **testosterone**
male hormone secreted by the testes, responsible for libido and secondary sexual characteristics.

▪ **tetanus**
a specific disease of the domesticated animals and

men, caused by *Clositridium tenani* which obtains access to the tissues through a wound. Horses are most commonly affected. The organism is present in most cultivated soil, specially such as receive heavy dressings of farmyard manure, in which it is extremely plentiful.

- **tetanty**

an animal disorder which is related to calcium deficiency, gastrointestinal disturbances etc. it is characterised by marked convulsions, rigid muscles, stiff legs ant] nervousness

- **tether**

to tie an animal with a rope or chain to allow gracing but prevent straying.

- **tetrad**

a unit of four chromatids formed as a result of synapses of homologous chromosomes, each of which consists of a pair of identical chromatids joined at the centromere.

- **texel**

a breed of sheep originating in Holland and widely distributed in Europe. It is known for its milk production and also noted for its rate of growth and potential for meat production particularly in France where a distinct meal type has been developed).

- **texture**

1. fabric—the closeness of the weave, structure and finish.
2. raw wool—the handle or feel of the wool,

- **therapeutic**

curative; referring to the art of healing. therapy. the medical treatment of disease.

- **therm**

the quantity of heat required to raise the temperature of 1,000 kilograms of water pc, or 13000 pounds of water 4f. One therm is 1,000 calories, or 1 mega calorie (meal).

- **thermal**

refers to heat.

- **thermal death point**

that temperature at which an organism is killed after an exposure of ten minutes under certain specified conditions.

- **thermal elements**
include temperature, humidity, air movement and radiant heat.

- **thermistor**
a small solid-state device that changes resistance markedly with changes in temperature, hence it is used to convert changes in temperature to changes in voltage of current.

- **thermocouple**
a device consisting essentially of two conductors made of different metals, joined at both ends, producing a loop in which an electric current will flow when there is a difference in temperature between the two junctions.

- **thermo genesis**
the chemical production of heat in the body.

- **thermolysis**
the joss or dissipation of body heat.

- **thermo neutrality**
the stare of thermal balance between an organism and its environment so that the body thermoregulatory mechanisms are inactive. the thermo neutral zone is also referred as the comfort zone.

- **thiamine**
Thiamine Hydrochloride, or vitamin $B_1$.

- **thigh**
that part of the leg above the hock.

- **thiouracil**
an anti-thyroid agent which lowers the rate of metabolism.

- **thirds**
a term once given to a category of wheat offal resulting from milling. All wheat offal, apart from bran, are now sold as weatings for use in feeding-stuffs.

- **thixotropic**
the property of certain gels, when mechanically agitated, to become fluid without a change in temperature. they reconvert to the gel slate when allowed to stand.

- **three way cross**
a system of breeding invoking the mating of males of three different breeds used in rotation on crossbred females.

- **threshold**
  the level or point at which a physiological effect become evident as a result of stimulation.

- **thrifty**
  a term applied to animals which are developing well, gaining weight, producing milk or eggs well, etc. and generally thriving,

- **thrombosis**
  the obstruction of a blood vessel by the formation of a blood clot.

- **throw**
  to cause an animal, as a horse or cow, to fall to the ground before branding, etc,

- **thumb mark**
  a depression on the side of a single comb (a defect).

- **thumps (thumping)**
  a respiratory disturbance in which the pigs breathe with difficulty, rapidly and spasmodically; jerky breathing.

- **trypsin**
  a proteolytic enzyme of the pancreatic secretion. It changes proteins into peptones and forms the main constituent of pancreatic extracts used for digestion of food. It is often helpful in cases of non-specific diarrhoea in dogs.

- **tryptophan**
  one of the essential amino acids.

- **tryptophan pyrrolase**
  an tryptophan to formyl kynure-nine. It is an iron porphyrin protein.

- **T sums**
  a system of Dutch origin, which purports to indicate the best lime to apply nitrogen to make the best use of grassland. It involves accumulating the daily average of maximum and minimum air temperature (°c) from 1st January until the total reaches 200, at which point application is carried out. Positive temperatures are added cumulatively whilst negative ones are ignored. The date at which T-sum 200 is reached varies in different parts of the country.

- **tub washed**
  wool that has been washed after shearing, a practice that has been almost entirely abandoned in north American, but formerly in

vigoue in Canada and some of the south-eastern states.

- **tuber**
  an enlargement or swelling.
- **tubercle**
  a packed in collection of epjthelioid cells admixed and surrounded by small round cells.
- **tuberculin**
  a biological agent derived from the growth and further processing of the tubercle bacilli which is used for detection or diagnosis of tuberculosis in animals and man.
- **tuberculosis**
  a contagious disease of many animals including man and all domesticated animals, mainly cattle and pigs, caused by the bacterium *Mycobacterium tuberculosis*. Characteristic tubercles or nodular swellings develop in one or more of the organs of the body and degenerate, becoming cheesy, so that affected organs are destroyed. T.B. of the lungs is the most common and is usually associated with coughing (chronic in cattle) and loss of condition.
- **tuberosity**
  a large rounded projection.
- **tubular heaters**
  tubes through which milk is pumped to pick up heat from stem, hot water, or on other healing medium that surrounds the tubes. Numerous designs are available.
- **tumbrill**
  a lip-cart used for carrying dung.
- **tumour**
  a new circumscribed growth, arising in parallel of normal rate of growth of pre-existing tissues.
- **tunnel drying**
  a method of drying bales of hay built into a stack with a central tunnel blocked at one end. So that air blown into it with a fan is forced up through the stack. The stack can be constructed in the field or more usually under a Dutch barn when it is also called barn hay drying.
- **tup**
  a ram.

- **turbidity test**
  a statutory test which must pass before it can be sold for public consumption as sterilised milk. The principle of the test is that when milk is boiled all the albumen in it is precipitated. It involves the addition of ammonium sulphate to precipitate other substances e.g. casein which are then filtered out the filtrate is heated and any albumen in the milk is revealed by turbidity. If the milk, had been adequately sterilised (by heating to in excess of 100°c) then the albumen would all have been previously precipitated and the test would reveal no turbidity.

- **turken**
  bare neck chicken; mistakenly held by some to be a cross between the chicken and the turkey.

- **turkey**
  a large poultry bird of the pheasant family native to North America, several varieties of which are kept for their meat. nowadays the main produce are specialists keeping large numbers of birds, although some small-scale producers still keep turkeys, themselves undertaking the killing, plucking and dressing of oven ready birds (mainly at Christmas).

- **turkey cock**
  an adult male turkey older than 26 weeks. It is also called *turkey stag, or tom.*

- **turkey grower**
  a turkey between 3 and 26 weeks of age.

- **turkey lion**
  a female turkey older than 26 weeks (turkey grower).

- **turn**
  a term used of a female animal which comes on heat again after being mated. Also called *return*.

- **turn up**
  to lift a sheep so that it sits on its rump to facilitate shearing, veterinary treatment, etc.

- **tush**
  a tooth located between the molars and incisors; the eye-tooth;

- **tusk**
  an elongated or greatly enlarged tooth.

- **twins**
homozygous identical twins, from the same ovum.

- **twitch**
to firmly squeeze the skin on the end of a horse's nose by means of a small rope that is twisted.

- **two crop ewe**
an ewe that has borne two crops of lambs.

- **type**
a model or ideal of perfection which combines all the characteristics that contribute to an animal's value and efficiency for the purpose specified.

- **type classification**
a programme sponsored by breed associations whereby a registered animal's conformation may be compared with the 'ideal , or 'true', type of animal of that breed by an official inspector (classifier).

- **typhus**
sensitive contagious disease of animal which is characterised by high temperature and great prostration.

- **typical**
presenting the distinct characteristics of the type. Tyrosine. One of the non-essential amino acids.

- **uberous**
yielding milk in abundance

- **udder**
the organ containing the mammary glands of animals. It is generally applied to the pendulous bag hanging beneath a cow just in front of the rear legs, bearing four teats. Also called bags.

- **udder clap**
mastitis.

- **ultra high temperature**
temperatures of 85° higher as applied to pasteurisation or sterility milk in a relatively short time.

- **ulcer**
a break in continuity of the epidermis with loss of substance and exposure of underlying tissue. it is slow to heal or tends not to heal at all.

- **ulceration**
growth of a condition whereby substance is lost from a cutaneous or mu-

cous surface, causing gradual disintegration and necrosis of the tissues.

- **ulna**

the lateral of the two bones of the forearm.

- **ultra heat treated milk**

a title of milk sold for liquid consumption after subjection to ultra high temperature to sterilise it by killing bacteria and other organisms, including spores, present in it. The milk is heated and held at a temperature of at least 270°F (132°c) for one second before cooling. This treatment destroys some vitamins and probably some proteins and affects the palatability of the milk. Before sale U.H.T. milk must pass the statutory colony count test to determine the possible presence of bacteria). It is sold in sterile packages, often cartons, to prevent recommendation and can be stored indefinitely if kept sterile,

- **ultra pasteurisation**

the process of heating milk at ultra high temperature for a sufficient time to kill all pathogenic micro-organisms, a temperature such as 88.3°c for 1 second.

- **undecylenate ointment**

a fungicide, used in the treatment of ringworm, etc.

- **under colour**

the fluff or part the feathers only seen when the plumage is raised. It is of greater value id breeding and frequently differs from the surface colour.

- **under shot (prognathia)**

a state involving inequality in the length of the jaws in which the lower jaw appears too long.

- **undulant fever**

a recurrent feverish disease of man which can last for a long period, occasionally many months. It is caused by the bacteria called *Bruiella abortus* which is responsible for *brucellosis* and is usually contracted by contact with infected cattle or by drinking un-pasteurised infected milk.

- **ungulates**

the term used for the herbivorous hoofed animals, e. g., cattle, deer, horses, pigs and sheep.

- **uniform price**
  the price paid producers for market milk when classified pricing is used in a marketwise pool. The uniform price is an average of class prices weighted by the quantity of milk in each class.
- **unilateral**
  that which affects only one side.
- **uniparous**
  producing only one offspring or one egg at a time.
- **unit**
  abbreviation for milking unit.
- **unkindly**
  a term applied
  1. to crops or animals which lack thrift or are in poor condition.
  2. to soils which lack natural fertility or are cold, late, wet, etc. and
- **un-merchantable**
  a tern that was at one time applied to wool that had been washed but was not sufficiency clean to he classified as standard washed wool. the term as used in this sense has become obsolete since the practice of washing wool has disappeared.
- **unsaturated fat**
  a fat formed from the reaction of glycerol with any one of several unsaturated fatty acids. Olein and linolein are examples.
- **unsaturated fatty acids**
  any one of several fatty acids containing one or more double bonds. Oleic, linoleic, linolenic-and arachidonic acids are examples.
- **unsex**
  to castrate a male or female.
- **unthriftiness**
  lack of strength, meagre growth or development; the quality or state of being unthrifty in animals.
- **untreated milk**
  milk which is sold for liquid consumption direct from farms as opposed to bring sent to dairies and without any heal treatment (heat treated milk),
- **unwashed wool**
  wool that is in its original, greasy state.
- **urea ($NH_2CONH_2$)**
  a crystalline, white sub-

stance found in urine, blood and lymph, which is the final product of protein metabolism in the body. synthetic urea can be used by ruminal bacteria as a nitrogen source for protein synthesis.

- **urease**
  an enzyme which acts on urea to produce carbon dioxide and ammonia. It is found m the jack bean and the soybean and is produced by certain micro-organisms in the rumen.

- **uremia**
  a toxic growth of urinary constituents in the blood.

- **ureter**
  It is the tube which carries the urine excreted by a kidney down to the urinary bladder.

- **urethra**
  the tube which leads from the neck of the bladder to the outside, opening at he extremity of the penis in the penis and into the posterior part of the

- **urogenital (genitourinary)**
  pertaining to the urinary and genital tracts (including the kidneys and sex organs).

- **urinometer**
  an instrument designed for the estimation of the specific gravity of urine.

- **uropygial gland**
  the preen gland (used by birds to water-proof their feathers).

- **uterine endometrium**
  the inmost layer of tissue within the uterine wall.

- **uterine involution**
  the loss in weight and length of the uterus after farrowing.

- **uterus**
  womb or female organ where the embryonic foetuses are nurtured.

- **utility**
  referring to a fowl bred for laying purposes.

- **utility stock**
  birds bred for commercial egg production, table purposes or a combination of eggs and meat.

- **vaccine**
  a suspension of disease producing micro-organisms modified by attenuation or by killing so that it will not

cause disease and can stimulate the formation of antibodies upon inoculation, i.e., vaccine is administered with the object of stimulating the recipient's specific defence mechanism in respect of given pathogen(s) or toxic agent(s).

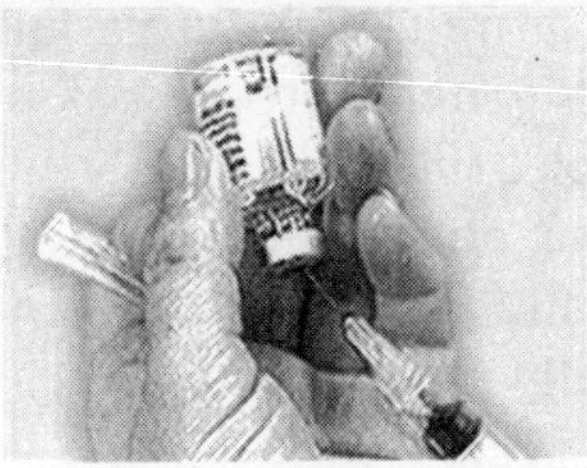

- **vacillation**
  the process of development of small empty spaces m cells or tissues.
- **vacuum**
  air within a confined space that is of lower pressure than atmospheric pressure.
- **vacuum atmosphere pulse**
  the alternating cycle pot out by a pulsator or relay.
- **vacuum controller**
  an apparatus to regulate a pre-determined level of vacuum within a confined space.
- **vacuum gates**
  gates which are opened and closed by vacuum acting on a piston in a cylinder.
- **vacuum gauge**
  a dial to indicate the vacuum level in an enclosed system.
- **vacuum level**
  the level at which vacuum is held by the controller.
- **vacuum line**
  a rigid tube that convey the vacuum generated by the pump to the milking plant,
- **vacuum phase**
  the phase in a pulsator or liner when the system is under vacuum.
- **vacuum pump**
  a device for reducing the pressure in an enclosed system below atmospheric pressure.
- **vacuum reserves**
  another name for air reserves.
- **vacuum silage**
  fodder prepared in a large, scaled, polythene 'bag'. The 'bag' is evacuated of air by a pump through a valve in the sheeting. This removal of air and the build up of

carbon-dioxide reduces respiration losses and aids fermentation, producing high qua lily silage.

- **vacuum stability**
  the capacity of a plant to hold the vacuum level steady whilst of varying demands of vacuum are being met.
- **vacuumisation**
  in milk processing, the subjection of heated milk to vacuum with the purpose of removing volatile off-flavours and odours and/or moisture.
- **vagina**
  the terminal portion of the hen's oviduct, through which the egg is expelled.
- **vaginitis**
  inflammation of the vagina.
- **valeric acid**
  one of the volatile fatly acids produced by ruminal bacteria.
- **valine**
  one of the essential amino acids.
- **vane**
  component of a vacuum pump. They fit into the rotor and by their action generate a vacuum.
- **vaporisation**
  the conversion of a solid or liquid into a vapour without chemical change.
- **variance**
  the clay of the breeder. Variance is a elastic that describes the variation we see in a trait. Without variation no genetic progress is possible, since genetically superior animals would not be distinguishable from genetically inferior ones.
- **variant**
  1. a specimen differing slightly in its characteristics from the type.
  2. a variable quantity whose measure men is or frequencies form all or part of the data for analysis.
- **variate**
  a variate is quantity which may take any of the values of a specified set with a specified relative frequency of probability.
- **variation**
  1. coefficient of .
  2. the standard deviation expressed as a percentage of

the mean of a population, it is a measure of relative dispersion.

- **variety**
  refers to a group of animals within a species or sub-species which share similar characteristics, but which differ in respect of those characteristics from other groups or varieties within the species. used loosely to mean a variation of any kind within the species, also called breed, race or strain.
- **varicocele**
  a condition in which the veins of one or both testicles are greatly distended.
- **vas**
  a duct or tube.
- **vas deferens**
  tube connecting the epjdidymis of the testicle to the urethra.
- **vascular**
  consisting of, pertaining to, or provided with vessels (such as blood vessels).
- **vasectomy**
  sterilisation of male animals by severance of the vas deferens. This blocks the passage of sperm to the outside, although sperm and male sex hormones are still produced.
- **vasoconstriction**
  reduction in the diameter of blood.
- **vessels**
  specially arterioles.
- **vasoconstrictor**
  a compound or substance that causes constriction of blood vessels.
- **vasodilation**
  the dilation of blood vessels resulting from stimulation by a nerve or drug.
- **vermicide**
  any material that kills internal parasitic worms.
- **vertebra**
  a single bone of the spinal column.
- **vertebral**
  pertaining to the vertebrae.
- **vertebrates**
  animals with backbones.
- **vesicle**
  a blister, or small sharply circumscribed elevation of the skin filled with clear fluid. Lesions larger than 0.5 cm

in diameter are called *bullae,* rarely seen in dogs.

- **vesicular**
  relating to or containing vesicles.
- **vestibule**
  space at the entrance of a canal; for example, at the vulva.
- **vetch**
  a leguminous plant, *Vicia saliva*, with a slender, square, climbing stem, tendrilled pinnate leaves and blue or purple flowers. It is sown with rye to provide spring keep for sheep and with oats, or beans and oats, for silage. It may also be made into hay. sometimes it is grown as a pure crop following cereals to produce seed. winter and spring varieties are available.
- **veterinary science**
  science and art of treating disease of animals. It encompasses anatomy, physiology, breeding, nutrition, animal diseases and treatment, man's use of animals, etc.
- **veterinary surgeon**
  a person qualified in veterinary science (the study of animal diseases), and able to diagnose diseases, prescribe medicines and carry out surgery.
- **viable**
  a term applied to a newly horn animal to indicate that it is capable of living separately from the dam. it is also used in connection with the survival of bacteria, etc.
- **viability**
  the ability to live.
- **VFA (volatile fatty acids)**
  usually used in reference to acetic, propionic and butyric acids produced in he rumen of cattle, goats and sheep, in the caecum of sheep, the caecum and colon of swine, the colon of the horse and the caecum of the rabbit.
- **villi**
  plural of *villus*.
- **villus**
  a minute vascular protrusion on the surface of a mucous membrane.
- **VIP**
  abbreviation for Verified Identification Programme.
- **viral**
  relating to viruses.

- **viremia**
 an infection of the bloodstream caused by a virus, virgin wool. Wool that has been used in the manufacture of fabrics for the first time. The term is used in distinction from all wool or pure wool which may include fibres recovered from rags.
- **virginiamycin**
 an antibiotic which may be included in livestock rations.
- **virion**
 a mature virus; the ultimate phase in viral development
- **viroid**
 name given to a particle of nucleic acid which is much smaller than a virus but lacks the tough protein cost of the virus. It is a disease causing particle in animals.
- **virology**
 the study of virus and virus disease.
- **virosis**
 a disease caused by a virus.
- **virucide**
 a chemical or physical agent that can kill or inactivate viruses; a disinfectant.
- **virus**
 one of a group of minute infectious agents. They are characterised by a lack of independent metabolism and by the ability to replicate only within living host cell . they include any of a group of disease-producing agents composed of protein and nucleic acid. viruses are filterable and cause such diseases in man as rabies, poliomyelitis, chicken pox and the common cold.

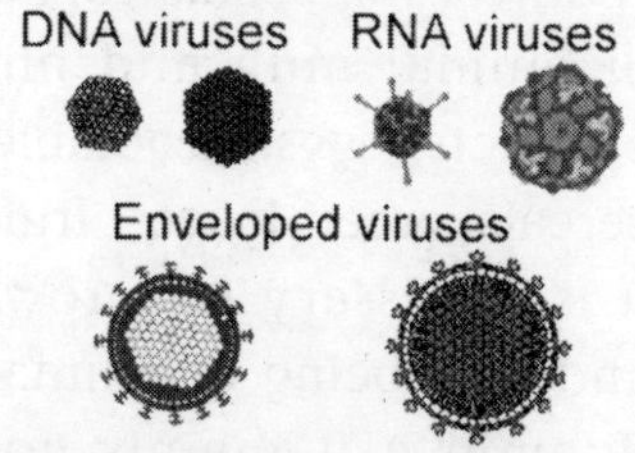


- **viscera**
 the internal organs of the body, particularly in the chest and abdominal cavities, such as the heart, lungs, liver, intestines and kidneys.
- **visceral**
 pertaining to the viscera.
- **viscosity**
 the freedom of flow of liquids.

- **visibility**

the longest distance in a given direction at which common objects like buildings or trees are visible to the unaided eye.

- **vitamer**

a sterol allied to cholesterol or ergosterol: a fat soluble vitamin. It is not destroyed by aeration but heat. Vitamin A is found in oils and fats, butler, halibut liver oil, cod liver nil, cellular organs of animal milk and milk products, eggs, vegetables, cereals, green leaves, fruits. it is necessary for growth and well being of animals. Vitamin A is mainly concerned with the maintenance of healthy epithelial structures.

- **vitamins**

a class of organic substances required by animals in small quantities, essential to metabolism and thought to act together with enzymes. Animals are able to synthesise several vitamins in their bodies but must be supplied

in their food. Vitamins A, D, E and K are fat-soluble and are stored in the fat compounds of the body. Vitamins B and C are water-soluble and cannot be stored and are regularly required in the diet. All farm animals are able to synthesise sufficient vitamin C in their bodies. ruminants and horses contain gut bacteria capable of producing sufficient supplies of the B group vitamins. Livestock kept outdoors are able to produce adequate supplies of vitamin D during the summer through the irradiation of their coats of skin.

- **viviparous**

producing living young (as opposed to eggs ) from within the body in the manner of nearly all mammals, many reptiles and few fishes.

■ **voids**
empty or unfilled spaces; in spray-dired milks, the open space or spaces within the particles.

■ **volar**
the sole surface of the fore-foot.

■ **volatile fatty acid**
any one of several volatile organic acids found especially in rumen content1 and/or silage. Acetic propionic and butyric acids are ordinarily the most prevalent,

■ **vomit**
1. to eject such material.
2. material ejected from the stomach from the mouth or nostrils.

■ **vomiting**
the forcible expulsion of the contents of the stomach through the mouth.

■ **waste food**
defined as any meat, bones, blot, offal, or other part of the carcase of any livestock, or of any poultry, or product derived there from, or hatchery waste or eggs or eggshells, or any broken or waste foodstuffs (including table or kitchen refuse, scraps or waste), which contain or have been in contact with any meat, bones, blood, offal, or with any other part of the carcase of any livestock or of any poultry.

■ **wasty**
the term applied to a carcase which has too much fat, requiring excessive trimming; may also be applied to paunchy live animals.

■ **wasty wool**
wool which is short, weak and tangled in the lock and as a consequence wastes m spinning and combing.

■ **wattles**
the red hanging structures at each side of the' bottom of the beat. Developed mostly in the male birds.

- **weeping**
a external inflammatory process in which there is considerable exudation of clear scrum from rupture of vesicles or discontinuity of the epidermis.

- **wean**
removal of a pig from its dam so that it can no longer nurse.

- **weaner**
a pig that has reached the age that it can be weaned or that has been weaned.

- **weaner**
a weaned piglet in the early growth stage.

- **weaning**
method of separation of suckling piglets from the sow.

- **web feather**
the hat part each side of the quill or shaft.

- **web of wing**
the triangular part of skin seed when wing is open.

- **westrain**
a cross-bred pig derived from a Belgian pie train and the old Wessex saddleback.

- **wet feeding**
the feeding to livestock of meal mixed with water in the form of a porridge.

- **wet fencing**
water-filled ditches and dykes in marsh areas which livestock will not or cannot cross and which therefore act like fences in retaining animal .

- **wet rendered**
cooked with steam under pressure in closed tanks.

- **wheat middling**
consists of fine particles of wheat bran, wheat shorts (a by-product of milling that consists of bran, germ and coarse meal) and some of the offal from the 'tail of the mill' and must contain not more than 9-5% crude fibre.

- **whinny**
the mild, soft cry of a horse.

- **whip tall**
the slim narrow tail, for example, of a modern game.

- **white clover**
a perennial type of clover *Trifolium repens*, characterised by a creeping growth habit which knits the sward together, keeps out weeds and unproductive grasses and minimises the destructive effect of peaching by livestock. Several varieties are available, all of which are persistent in the sward out are less bulky than red clover, white clovers tolerate grazing well and are used in all long leys. (Dutch White Clover and Wild White Clover).

- **white corpuscles**
leucocytes of the blood.

- **white straw crops**
a term used for wheat, barley, oats and rye.

- **WHO**
World Health Organisation. An agency of the United Nations founded in 1948, it seeks to promote worldwide health and prevent outbreak of disease. It assists countries in strengthening public health services. It plans and co-ordinates inter-national efforts to solve health problems, with special attention, to malaria, tuberculosis and venereal virus and parasitic diseases. It works with member countries and other health organisation to collect information on epidemics; develop international quarantine regulations; and standardise medical drugs, vaccines and treatment. About 100 countries belong to WHO, which headquar-

tered is in Geneva, Switzerland.

- **whole**
  a term used to mean complete,

- **whole body counter**
  an apparatus used to identify and measure the radiation in the body of man and animals; it used heavy shielding (to keep out background radiation), ultra sensitive scintillation detectors and electronic equipment,

- **whole milk**
  milk as it comes directly from the udder with nothing removed.

- **whorl**
  a twirl, or cowlick, in an animal's hair.

- **wild white clover**
  a small-leaved, vigorous, truly perennial form of white clover, the stolons ofuhich spread close to the ground keeping out weeds. It is drought resistant, persistent and remains continuously palatable to grazing animals.

- **wiltshire side**
  the entire half of a dressed pig, minus the head, shank, shoulder bone and hip bone. all of the side, except the ham and shoulder, is sold as *'bacon'*.

- **wind gall**
  a bulging swelling in the area of a horse's fetlock.

- **windrow**
  a row of hay (or other cut crops), consisting of two or more swaths combined into one in preparation for picking up.

- **wins bar**
  any dark line of colour or marking across the middle of the wing caused by the lower wing coverts.

- **wing bay**
  the three concerned part of the folded wing below the bar.

- **wing bow**
  the shoulder parts of the wing above the bar.

- **wing butts**
  the ends or corners of the

wings. The upper ends being more properly the shoulder butts, the lower ends are termed the lower butts.

- **wing coverts**
  the broad feathers covering the roots of secondary quills.
- **wing shoulder**
  a state in which the moulder joint is away from the rib structure and skeleton (much of the skin is tucked in behind it).

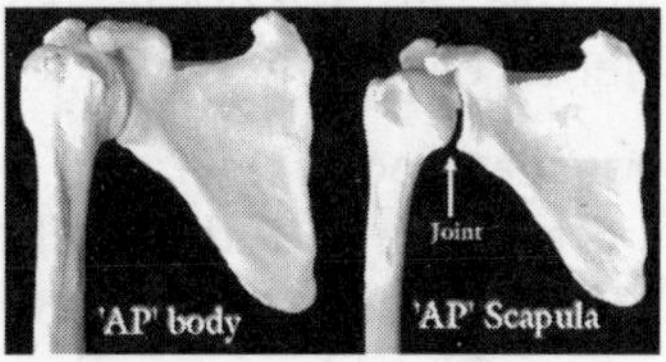


- **winter annual**
  a plant species which initiates growth in the fall, lives over winter and dies after producing seeds the following season.
- **winter feeding**
  the feeding regime of livestock, mainly cattle, in the winter months, involving the feeding of hay, silage, other bulk foods and concentrates.

- **winter green**
  a term applied to plants which remain green and fresh through the winter.
- **winter pause**
  a break in egg production in the winter period which is usually accompanied by a partial moult.
- **wish bone**
  clavicle bone in form of breast bone.
- **wool ball**
  a mass of wool, tangled into ball found in the first or fourth stomach of lambs. Wool is swallowed from the mothers' fleece when suckling and accumulates, sometimes the stomach outlet and causing death.

- **wool fat or wool grease**

  the waxy, grassy substance that exudes from the sheep skin and clings to the wool fibres. In the crude form it is often known in commerce as Degras. After refining, it is sold in the drug trade as *lanolin*. It is soluble in ether, but not soluble in water.

- **wool rot**

  condition of sheep due to a fungus which causes irritation to the skin in wet weather and the development of hard, yellowish scabs. The sores heal and the rowing wool carries the scabs away in the fleece. also called lumpy wool.

- **wool sorter or wool stapler**

  one who sorts wool.

- **wool sorters disease**

  an anthrax, very often contracted by sorters and others who work in wool from sheep that have died of that disease. In this respect the most dangerous wools come from Eastern Europe and Asia.

- **wool sorting**

  the procedure of separating fleeces of wool into parts, according to fineness and length. This process should be distinguished from grading in which process the fleeces are not divided.

- **woollen system**

  a method of yarn production from wool fibres which have been carded, but not combed or gilled. the fibres in the yarn do not lie parallel to each other.

- **woollen wools**

  a term applied in short wools that are adapted to manufacture of woollen as distinguished from worsted fabrics.

- **woollens or woollen fabrics**

  fabrics conducted in whole or part from woollen yarn, i.e., yarn made from uncombed wool.

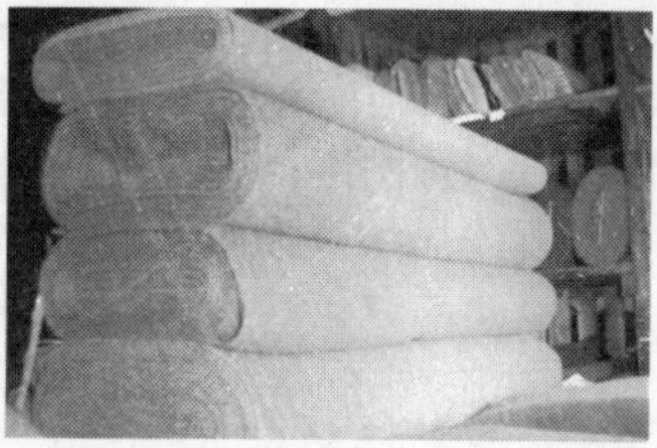

■ **work**
a term commonly associated with the use of horses to round up, cut. etc., cattle.

■ **worrying**
the tormenting or harassing of livestock, usually by uncontrolled dogs. this often involves the biting and tearing of flesh. sheep, which become particularly anxious when worried, are frequently tilled by dogs and pregnant ewes often about. farmers arc entitled to shoot dugs found worrying livestock

■ **worsted system**
a system of yam production from wool fibres which have been carded or prepared and either gilled or combed, or both. The fibres in the yarn lie parallel.

■ **wrinkles**
small folds of skin an the body of the sheep and over the breech region.

■ **wry neck**
neck carried down or to one side.

■ **wry tail**
a tail carried away to the right or left of the continuation of the backbone.

■ **xerophthalmia**
a disease of the eye associated with a vitamin a deficiency. Blindness may be produced.

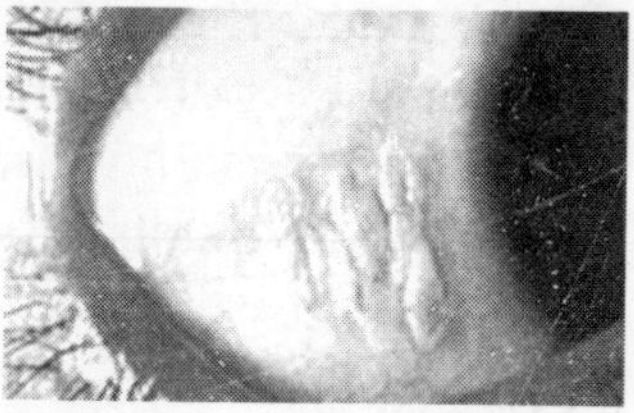

■ **xylazine**
the dynamic ingredient of '*Rompun*' , a tranquillising

used to render farm livestock easier to handle.

■ **X-rays**
radiation produced when electrons in a vacuum tube are projected at very high tension and velocity to strike a metallic target these are electromagnetic waves, by their wavelength is only about one-thousandth of that of visible light.

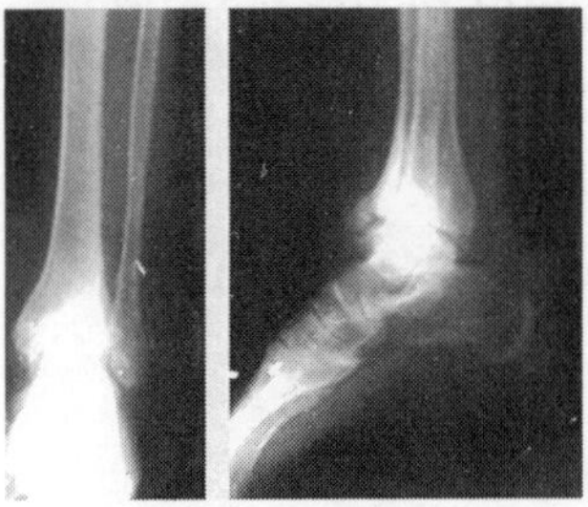

Before Total Ankle Replacement

■ **x, xx, xxx**
three grades of washed wool from merino sheep that were at one time made in the eastern part of the united states. The more X's the finer the wool. after the practice of washing disappeared in this country, these grades out of existence.

■ **x, y chromosome**
chromosomes concerned especially with the determination of sex. In acme animals the females have two X chromosomes and the males one X, accompanied by its unmatched mate, the Y chromosome. certain others, moths for example, have XX males and XY females.

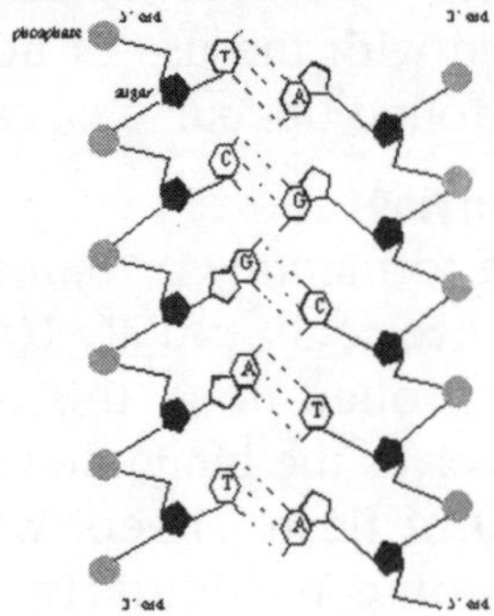


■ **yarn**
wool fibres spun into a thread.

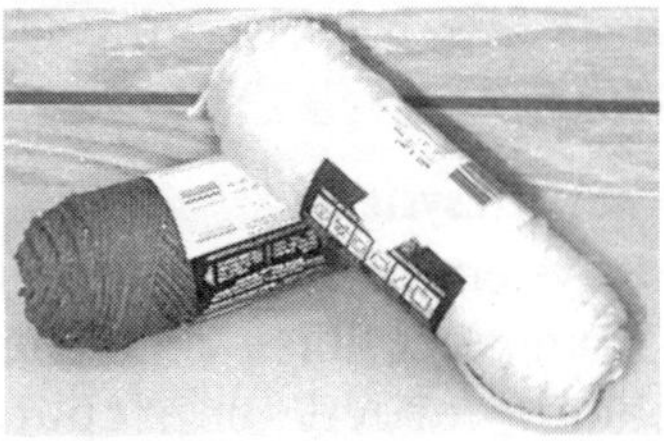

■ **yean**
to give birth to young, particularly by goats and sheep.

■ **yearling**
refers to a male or female farm animal (particularly cattle and horses) during the

first year of its life.

- **yeld mare**
  a dry (non-lactating) mare or a mare that has not raised a foal during a particular season.

- **yelt**
  a female pig intended for breeding, up to the time that she has her first litter.

- **yield**
  1. percentage of clean wool after scouring.
  2. dressing per cent of an animal slaughtered.
  3. pounds of wool clipped per ewe.
  4. volume of milk produced per day or per lactation.

- **yoghurt**
  fermented milk, low-fat milk, or skim milk, sometimes protein-fortified. milk-solid content is generally 1.5 per cent. most yoghurt is high in protein and low in calories.

- **yoke**
  1. a collar placed round the neck to keep it in a stall
  2. a frame joining a pair of draught oxen. Also any pair of animals working together.

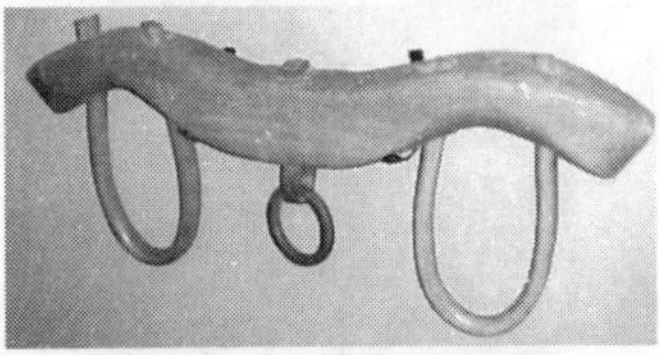

- **yolk**
  the natural exudations of the skin that adhere to the wool fibre. yolk is made up of
  1. suint, the water-soluble constituent; and
  2. wool grease, the ether-soluble constituent.

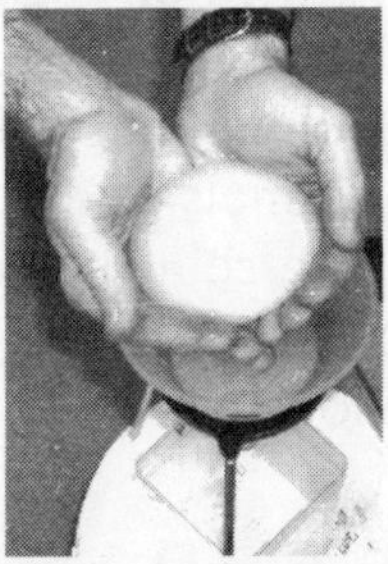

■ **yolk slain**
a stain caused by pigment from the yolk.

■ **Yorkshire fog**
a permanent, spongy, downy grass common in poor pastures and an indicator of lime deficiency in the soil. Also called meadow soft grass.

■ **zanil**
a drug used against liver-flukes in cattle and sheep.

■ **zebu**
this refers to a strain of cattle originating in India; widely domesticated throughout India, China and East Africa, used as beasts of burden and meat animals and for their milk. The zebu has a large hump over the shoulders, pendulous ears and a large dewlap. This is also known as *brahman*.

■ **zero grazing**
to supply cut fodder to yarded cattle, or to cattle in use paddocks. It has a place on heavy land with high stocking rates and large herds. Zero grazing obviates poaching and the spoiling of grass and a given acreage zero-grazed can provide more grass than if grazed. Cutting grass every day and mechanical failures can upset the system, it is not yet considered economic for sheep.

■ **zo**
it is a prefix which refers to animals.

- **zoonoses**

  these are a group of infectious diseases transmissible from animals to man. The most common among them may include salmonellosis, leptospirosis, campylobacter infection and Q-fever.

- **zoophilic**

  having a preference for animals. Used, in dermatological sense, to refer to zoophilic fungi whose primary host is an animal.

- **zootechny**

  animal management.

- **zootoxin**

  poison or toxin produced by animals, as snake venom or spider poison.

- **zooxanthin**

  yellow pigment found in animals.

- **ZPG (Zero Population Growth)**

  the point at which births deaths and the population stabilises, number wise.

- **zygoma**

  the bridge of bone which runs from near the base of the ear to the lower posterior part of the eye-socket,

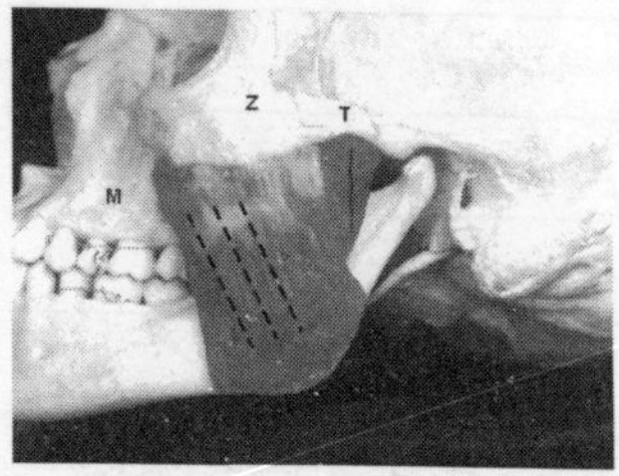


- **zygote**

  Zygote is a cell which is derived from the union of the

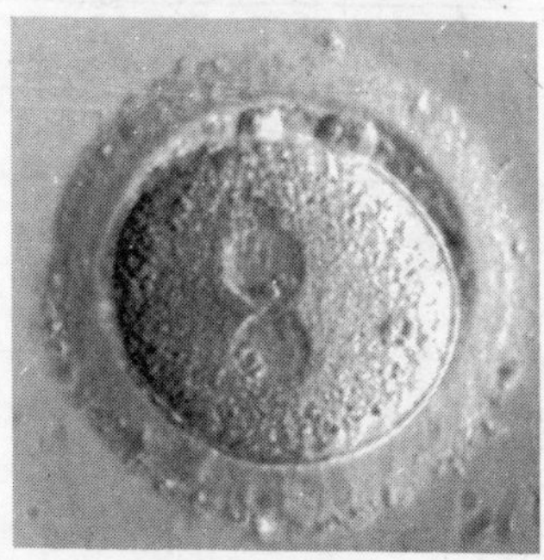

male and female gametes. This union restores the diploid number of chromosomes in the new cell.

# Notes